W9-CIP-596

TURNING OFF THE HEAT

TURNING OFF THE HEAT

WHY AMERICA MUST DOUBLE ENERGY EFFICIENCY TO SAVE MONEY AND REDUCE GLOBAL WARMING

THOMAS R. CASTEN

Prometheus Books

59 John Glenn Drive
Amherst, New York 14228-2197

Published 1998 by Prometheus Books

02 01 00 99   5 4 3 2

Library of Congress Cataloging-in-Publication Data

Casten, Thomas R.
    Turning off the heat : why America must double energy efficiency to save money and reduce global warming / Thomas R. Casten.
          p.   cm.
    Includes bibliographical references and index.
    ISBN 1–57392–269–2 (cloth : alk. paper)
    1. Energy conservation—United States. 2. Fossil fuel power plants—Environmental aspects—United States. 3. Global warming. I. Title.
TJ163.4.U6C38   1998
333.7916'0973—dc21   98–28272
                                                                               CIP

Printed by RR Donnelley & Sons in the United States of America on acid-free paper

Contents

Foreword
   Hon. Federico Peña   xi

Preface   1

1. ESTABLISHING THE PROBLEM
   ## AND A MARKET PERSPECTIVE   **15**

Global-Warming Concerns   15

Climate-Change History   18

Population Growth   21

Government Actions   22

2. UNDERSTANDING
   ## MONOPOLY BEHAVIOR   **25**

What Is "Monopoly"?   25

A Brief History of Electric Monopoly Protection   32

v

Diffusion of Innovations Takes Time   33
A Whale (Oil) of an Example   36
Problems of Early Electric Entrepreneurs   37
Problem Solved—Exclusive Franchises to Sell Electricity   38
Controlling the Monopolist's Greed and Politician's Income   39
Samuel Insull Invents State-Granted Monopolies   40
Economies of Scale   42
Generation Technology Primer   45
The Eighty-Billion-Dollar Monopoly Efficiency Gap   53
More Cooperation, Fewer Innovations   53

All Monopolies Are Slow to Innovate   55
You Get What You Reward   55

**"Natural-Monopoly" Characteristics of
    Electric Generation Are Gone   56**

PUHCA Protects Bad Managers   57

PURPA Induces Management Changes   59

Progress in Technology   63
Diesel-Engine Progress   63

Gas-Turbine Progress   65

Monopoly Protection Leads to Excess Electric Use   67

**3. NOBLESSE OBLIGE—
   FOR PLANET EARTH   70**

On Becoming God-Like   70

Too Rapid Energy Release   74

We Have Growing Environmental Problems   75

Fear of Economic Disruption Colors Policy   77

Progress in Ethics?   79

A Proven Way to Increase Human Thinking Power   80

4. SENDING PRICE SIGNALS   84

Unleashing Market Forces   84

Industry Is Energy-Conscious   85

Consumers Are Not Energy-Conscious   86

**Why Aren't Energy-Saving Technologies
Deployed More Rapidly?**   90
    Markets Depend on More than Price   90
    Energy Efficiency Is Proven High Return and Low Risk   92
    Problems with Landlord Investment in Energy Conservation   94
    Asking Utilities to Invest in Conservation—
        Demand Side Management   96
    Cultural Values Impact Decisions About Energy Efficiency   98

Changing American Values   101

5. LINKING DEREGULATION AND
    CARBON DIOXIDE REDUCTION   105

**The U.K. Experience Proves that Deregulation Saves
    Carbon Dioxide and Money**   105

A Recipe for National Action   107
    End all Restrictions on Electric Generation, Distribution, and Sales   107
    How to Double Electric-Generation Efficiency—
        Adopt a Fossil Fuel Efficiency Standard   109
    Set Pollution Limits Per Unit of Output for All Electric Producers   111
    Provide Transition Help to Monopolists   112
    Summary of Above Principles   113

**Utility Stranded-Asset Claims Prove
    Econometric Models Are Wrong**   115

A Closer Look at Pollution Reduction   117

6. FACING THE CHALLENGE: STABILIZING GLOBAL CARBON DIOXIDE   119

Stating the Real Problem   119

Technology Should Not Be the Main Focus   123

We Need Details Before Adopting the
     Kyoto Protocol Targets   124

Causes of Carbon Dioxide Emissions   126

Changing the Rules to Lower Carbon Dioxide
     by 60 Percent—Sketches of a New Paradigm   127

7. CHANGING CONSUMER ATTITUDES TOWARD ENERGY USE   130

New Building Codes   132

Tax Credit for Energy-Efficient Appliance Purchase   133

Encourage District Energy System Construction   134

Add an Energy-Efficiency Criterion to Federal Support   138

End All Fossil Fuel Subsidies   140

National Shade-Tree-Planting Program   143

Tax Incentives for Fuel-Efficient Cars   144

8. BARRIERS TO EFFICIENCY   146

"Why Aren't These More Efficient Energy Technologies
     More Widely Used?"   146

Understanding Barriers to Efficiency   148
     Technology Changes in the Mid-1960s   150
     Today's Technologies Are Even Better   150
     Monopoly Rules Slow Adoption of New Technology:
         You Get What You Reward   153
     Excessive Rules Resulted in Ever-Larger Central Plants   154

Small-Scale Power Generation Improves   155
Inefficient Fuel Use Creates Problems   157

Barriers to Efficiency   157
Laws and Regulations to Protect Monopoly   158
Regulated Monopolies Have No Incentives to Be Efficient   159
Retail Electric Sales by Competitors Prohibited   162
Laws Ban Transmission Competition   164
Interconnection Rules Controlled by Competitor   165
Backup Power from Competitor   169
Condoned Anticompetitive Practices of Monopolists   172
Bundling Heat and Power   172
Buying Out Cogenerators   173
Real-Time Pricing   175
Percent of Load Rate Ratchets   176
Environmental Laws Ignore Efficiency or Assume Central Generation   177
Input-Based Rules Ignore Efficiency   177
The Present Regulatory Approach Extends Use of Inefficient Plants   179
Air Permits Required to Start Construction Delay Efficient Plants   182
Allow Pollution Offsets   184
Regulations Prohibit Optimizing Plant Operations with the Grid   185
Tax Policy Penalizes Efficiency   187
Federal Reimbursement Policies Do Not Reward Efficiency   188
State Laws Based on Obsolete Technology Prevent Efficiency   190
Example #1: State Law Blocking Efficiency   190
Example #2: Making an Example of MIT   191
Example #3: State Law Blocking Efficiency   193
Example #4: State Law Blocking Efficiency   193
Example #5: State Law Blocking Efficiency   194

Much Energy Can Be Saved by Rule Changes   195
Include Barriers to Efficiency in Economic Models of
Climate-Change Mitigation Costs   195

9. FEDERAL ACTION TO DECARBONIZE   198
Finding One New Signal that Affects All Energy Thinking   198

Forces Blocking Deregulation of Electricity   201
States' Rights Block National Action   201
Rural Electrification Act   202
Municipal Power Companies   202

Federal Power Agencies   203
Utilities as Tax Collectors   204

How to Change Power-Producer Behavior?   205

Three Current Federal Proposals to Decarbonize   206
Carbon Cap—Impact and Problems   207
Revenue-Neutral Taxes on Fossil Fuel or Energy   209
Renewable Energy Portfolio   211

**Economy Versus Environment—
Finding a Middle Ground**   212

A Fossil Fuel Efficiency Standard   213

**Can A Fossil Fuel Efficiency Standard Reduce
Carbon Dioxide and Save Money?**   216
Rationale for a Federal Fossil Fuel Efficiency Standard   220
Avoid a Federal Fuel Choice   221
Fossil Fuel Efficiency Is Essential to a Sustainable Future   222
A Fossil Fuel Efficiency Standard Is Equitable and Predictable   223
Predicted CO_2 Reduction from Competition and a
    Fossil Fuel Efficiency Standard   224
Fossil Fuel Efficiency Standard Versus the Alternatives   227

Expected Electric Cost Reductions   228

**10. ERRATA—ENERGY REGULATORY
REFORM AND TAX ACT**   230

EPILOGUE   235

NOTES   239

GLOSSARY   247

INDEX   267

Foreword

Hon. Federico Peña,
Former United States Secretary of Energy and Secretary of Transportation

This book, authored by one of America's leading energy entre-preneurs, makes the case that America will find solutions to the climate-change problem in actions that will strengthen our economy. *Turning Off the Heat* presents a clear and sensible strategy showing how our country can simultaneously reduce emissions of climate-disrupting greenhouse gases and help consumers pay less for energy by modernizing our system for generating and delivering electricity.

Turning Off the Heat stands in that uncommon category of revealing books on government policy written not by an insider, academic, or advocate, but by a successful businessperson sharing decades of street-tested practical experience. Casten seeks to open our eyes to the fact that the way our country regulates, generates, and uses electricity is out-of-date and woefully inefficient. Drawing on his experience as an energy supplier throughout the United States, Casten argues that new energy technologies and energy service strategies, if allowed to work, could dra-matically reduce harmful emissions, including greenhouse gases, and trigger price reductions for energy consumers.

There is plenty of evidence to support Casten's thesis. Among the most telling is the disturbing fact that U.S. electricity producers typically use only one-third of the energy value, the BTUs, in the fuel they burn to create electricity. The rest of the energy, two-thirds of the total, is discarded as waste heat. This inefficient use of fuel and associated excess emissions of carbon dioxide and other pollutants occurs despite the fact that technology exists to capture and use more than 90 percent of the energy value of virtually any fuel, including coal.

Casten's core message is that an orchard of low-hanging fruit is out there ready for harvest. He calls on federal and state governments to take a top-to-bottom look at all the laws and policies that stand in the way of a clean, affordable modern energy system—and clear out the entangling underbrush. Casten believes that a modern regulatory scheme for energy production, including competition in energy markets, consciously aimed at encouraging America's energy entrepreneurs, will allow clean, economic energy sources to enter and ultimately dominate our energy system.

One does not have to agree with every idea in this book to recognize that Casten's thesis is both hopeful and inspiring. We know for a fact that American entrepreneurs are extraordinarily good at bringing new technologies to market. We know that Americans care about the environment and want the option of buying products and services produced in environmentally sustainable ways. We know that solar panels, fuel cells, and other clean technologies and techniques work and can be brought to market. We know that wasting two-thirds of the energy in fuels used to make our electricity is environmentally inexcusable and economically senseless. We know from our experiences that many laws and rules outlive their usefulness and produce unintended adverse consequences. Casten simply asks us to act on what we know.

During my service with the Clinton-Gore administration, we took several key steps in this direction. The administration's climate-change mitigation strategy embraces the view that electricity generation will become more efficient and less polluting if electricity suppliers are allowed to compete, and we proposed legislation to Congress that would stimulate such competition. We also proposed new tax and technology measures to boost non-emitting energy technologies. Though Congress has yet to act on the administration's proposals, I believe that a majority of the Congress will soon come to agree that these are changes our country needs.

How can we modernize our energy system to make it fit the economic and environmental priorities of the twenty-first century? *Turning Off the Heat* offers answers that deserve our attention and action. In the best traditions of our democracy, an informed citizen, Tom Casten, has stepped forward to offer sensible ideas based in experience that can bring people together and help move the country ahead. This is an important book for our times.

Preface

This book is about unleashing market forces to reduce human-kind's pollution of the atmosphere with by-products from burning fossil fuel. By ending electric-monopoly protection, eliminating barriers to efficiency, and guiding the energy industry to decrease use of fossil fuel, the United States can reduce emissions that are causing the earth to warm. These actions will also reduce the pollution that is producing acid rain and killing trees all over the world. This book explains why the average U.S. electric production is less than half as efficient as is economically achievable using present technology and how government officials can lead our society to a sustainable future. We have discovered the secret word—*competition*.

A United Nations panel of twenty-five hundred climate scientists concluded in 1995 that humans are having a discernible impact on the global climate by burning fossil fuel. By rapidly burning fossil fuel and releasing the carbon stored in the past four hundred million years, we are increasing the atmospheric concentration of carbon dioxide. Carbon dioxide and other gases emitted by human activity increase the global atmosphere's ability to trap heat and thus increase average global temperatures.

This book treats human-induced global climate change and the dying of U.S. forests from acid rain as real, but leaves to others detailed scien-

1

tific explanation behind both of these impacts of excessive fuel burning. The carbon dioxide concentration in the atmosphere has risen more than 25 percent in the last 150 years due to human activity, and unless the world stops burning so much fossil fuel, the atmospheric concentration of carbon dioxide will double in the next one hundred years. Higher carbon dioxide concentrations in the atmosphere mean a stronger greenhouse effect that will increase the heat the earth retains from the sun. Climatologists are concerned that a small increase in average global temperatures could produce other feedback effects that lead to rapid changes in climate. Heat differences drive global weather; higher average temperatures are predicted to increase the intensity and frequency of storms and unusual weather, including, paradoxically, extreme cold in some regions. The extreme weather events that we continue to experience, such as the El Niño storms in the 1997–98 winter, do not prove that climatologists' theories are correct, but what we are experiencing fits the predictions.

Others can debate the science and argue about the accuracy of the climate models. I think that waiting to become "absolutely positively certain" about the effects of increasing the carbon dioxide concentration in the atmosphere will create difficulties. Both well-established science and common sense tell us that we are creating a problem for future generations. We are playing an extremely high-stakes game.

We can reduce the amount of fossil fuel burned in the United States and the world without lowering our standard of living and without reducing the amount of energy we use. Fuel burning and energy use are not necessarily connected. We can burn less fuel more efficiently to produce the same amount of energy, waste less energy to avoid needless fuel burning, and/or make more use of renewable energy.

This book will not greatly increase your knowledge of the science of global warming or greenhouse gases, and does not attempt to duplicate the work done elsewhere on defining the problem. I recommend Ross Gelbspan's excellent book, *The Heat Is On* (Addison Wesley, 1997). Ross covers much of the science and describes the modeling of climate as well as the many weather events that fit what the models are predicting. Climatologists construct complex computer models that simulate weather systems and try to predict what actual climate changes will occur from such changes as increasing CO_2 concentration, increase in cloud cover, erupting volcanoes, or countless other changes that impact global climate. The book is rich with references to other documents about the science of climate change, and will lead to a better understanding of the topic. It does not do as good a job, in my view, of explaining what we can do to mitigate climate change. That is our task.

We do not explain the problem of forest dieback, the phenomenon of trees prematurely dying throughout a local forest. Charles E. Little's *The Dying of the Trees* (Penguin Books, 1997) pulls together many of the problems caused by acid rain from power plants, but Little does not explain how we could reduce acid rain and save money. The senseless burning of twice as much fuel as is needed to produce electricity and heat is the major reason for acid rain and other pollution that cause trees to die. Dying trees are simply another cost of protecting the electric industry from competition and from tolerating the continuance of barriers to efficiency.

I am not a scientist or a science writer. I am an entrepreneur who became concerned about human-induced carbon dioxide emissions in 1975 and have been building a business based on energy efficiency ever since. In some far-reaching studies I was performing for senior executives of Cummins Engine Company in 1974–75, global pollution was one of the main topics—were we destroying the planet? After much literature review and talking with the many scientists and engineers working in the Cummins Tech Center, I became convinced that new technology would be developed and deployed to solve most of the problems of human pollution. Evidence from the ensuing years has strengthened that view.

Control technologies that limit emission of oxides of nitrogen and sulfur have greatly improved and have become less expensive to install and operate. Advances in emission control have been made for every method of converting fuel to useful energy including diesel and gasoline piston engines, combustion turbines, boilers, and other heaters. New combustor technology is enabling gas turbines to produce only 2 to 5 parts per million of oxides of nitrogen, almost too little to measure. Contrast this to the 200 parts per million of nitrous oxide (NOx) emissions from the 1975 vintage combustion turbine and you find up to a 99 percent decrease in emissions. Yet two-thirds of all U.S. power is generated by plants that are over twenty-five years old and up to twenty times more polluting than plants using today's technology.

I also became convinced that technology would not be developed or deployed to control or remove carbon dioxide emissions from burning fossil fuel. There are no known theories for removing carbon dioxide without using so much energy that more carbon dioxide is created than is removed. *The only way to reduce carbon dioxide emissions is to burn less fossil fuel.* There are many proven ways to convert and use energy more efficiently that lower the price of energy to all consumers.

It is usual to assume our free markets work well and that we are optimally converting fuel to heat and electricity, given available technology. However, our energy markets are not free, and have not been free for seventy to ninety years (state regulation developed over a twenty-year period). All production, transmission, distribution, and sales of electricity are subject to monopoly protection, and regulation has been substituted for market forces. Regulation does a very poor job of optimizing the production and delivery of goods and services as proven by the miserable economic performance of every centrally planned economy in the world. Regulation of the electric markets has blocked innovation, erected countless barriers to efficiency, and removed the incentive for energy professionals to sell energy-conservation technology. This book explains how government can fix this problem by unleashing market forces to make such efficiency investments, and in the process mitigate climate change and save money.

The data for electric-generation efficiency prove the utter failure of monopoly regulation. The U.S. average delivered efficiency—the measure of electric energy delivered to the customer divided by the total fuel energy consumed—peaked at 33 percent in 1959 and has not increased since despite an incredible increase in knowledge and improvement in technology. *On average, two-thirds of all fuel burned to produce electricity in the United States is wasted.* Technology has made enormous strides and the best new electric-only generation plants that were built in 1997 have delivered efficiency of 57 percent, but our regulatory and environmental policies combine to keep old, inefficient generation plants in service. Two-thirds of the United States electric generating capacity was built more than twenty-five years ago, before the passage of the Clean Air Act. Besides being twenty to one hundred times more polluting than the newest plants, this old generation is not competitive and would be either greatly upgraded or shut down in a competitive world.

This exclusion of market forces from the electric arena also causes the United States to burn needless fuel to produce thermal energy—heat for our homes and offices and steam for industrial processes. Whenever any fuel is used to produce electricity, using any technology known, significant quantities of heat must be rejected. When appropriately sized generating plants are built near users of heat, the normally wasted heat from electric generation can be recovered and sold, thus saving money and avoiding burning more fuel to make heat. The process is called "combined heat and power" (CHP). It is not a new idea and was in fact utilized by Thomas Edison's very first commercial electric-generating plant on Pearl Street in Manhattan in 1881. Using the latest technology, a new combined heat-and-power plant can deliver electricity and heat with up to 91 percent efficiency—nearly three times the national average—and save money. With appropriate policies, we will rebuild our electric infrastructure, use the fuel savings to pay for the capital, and reduce electric rates by 30 to 40 percent.

The production of electricity accounts for one-third of all U.S. carbon dioxide emissions, and the energy waste is colossal. The energy thrown away by U.S. electric generators each year exceeds the total energy use of Japan. Then, we burn more fuel to separately produce our thermal energy. These two sectors of the United States economy—heat and power production—consume two-thirds of all the fuel burned in the United States. The remaining third is consumed by transportation. Addressing the waste in electric generation and capturing and using the waste heat from electric generation to replace fuel burning for thermal energy addresses both sectors. By unleashing competition in the electric markets, we will cause the energy industry to burn half as much fuel, produce half as much carbon dioxide, and reduce other pollutants by 95 percent or more.

What is surprising is that these changes reduce energy bills and make the economy more competitive. Those writing and speaking about reducing pollution almost always assume that energy use and fuel use are tied together. They say things like, "To reduce carbon dioxide emissions to the levels agreed to in Kyoto, Americans will have to reduce their energy use by 30 percent." In fact, what Americans must do is reduce the fuel used to produce energy by 30 percent, and this will save money. Burning less fuel to produce the same quantity of useful energy lowers the production of carbon dioxide. Burning less fuel saves money. This makes carbon dioxide a unique pollutant. It cannot be controlled by conventional pollution-control technologies, but is reduced by improving efficiency or by using less energy. Using less energy could adversely affect the economy, but improving efficiency saves money.

We also use too much energy. Energy waste is the direct but unintended result of our rules, regulations, and tax policies. By granting monopoly status to the electric industry, we have removed the incentive a competitive industry would have to help its customers conserve energy use. For example, if many energy firms compete to supply electricity and heat to a large office building, one or more of these firms will offer sophisticated controls and efficient lights and motors that conserve energy as part of their package. The building will end up with all necessary lighting, power, and comfortable temperatures, but will consume less electricity and heat.

We waste energy because the price is artificially low—at least the cost the user sees and pays is low. The United States and many other countries provide massive subsidies to fossil fuel, masking the true cost of energy and encouraging overuse. Overall, these subsidies do not help the consumer, because taxes are then collected to pay the cost of the fuel subsidies. The unintended consequence of choosing direct energy subsidies as a way to help people in need or to help domestic industry is energy waste. Since the individuals and businesses do not see or pay for the full cost of energy, they use more energy. They have less incentive to conserve, because the subsidized energy price is low. This generates more demand for fossil fuel. Ironically, the energy waste caused by the subsidies adds to fossil fuel demand and thus pushes up the world prices for fuel.

These subsidies also color everyone's thinking about the economics of using renewable energy resources—energy from the sun, wind, hydropower, or from burning biomass. Biomass is any organic substance, such as wood, straw, rice hulls, or other material produced by plants. All biomass contains stored energy derived from sunlight. Advances in technology and manufacturing efficiency have made several renewable energy technologies competitive with the true cost of fossil fuel, but many people think renewable energy is not competitive because they compare its cost with the price of subsidized fossil fuel-based energy. We will repeatedly point out the truism that, "You get what you reward," and we reward energy production with fossil fuel.

The best way to decarbonize the economy is to unleash market forces, to stop all government subsidies of fossil fuel, and to eliminate the many barriers to efficiency. This is easier said than done. There are a plethora of national, state, and local rules and a paucity of economists, environmentalists, policy makers, and elected officials who understand that unleashed market forces will increase fuel and end-use energy efficiency. Although market-induced efficiency would significantly mitigate global climate change and save money, the widespread view is that we already make heat and power as efficiently as possible. This myth is promoted in part by monopolists intent on retaining their monopoly rents. It is unsupported by the facts.

It will be difficult to quickly change all of the laws and regulations that are unintended barriers to efficiency. One out of four people has his or her electric power subsidized by the federal government, and nearly all of the thirty-two hundred separate private and public monopoly utilities resist losing monopoly protection. States exercise police powers to protect their "native" electric industries and resist federal control or loss of their ability to control the electric business. In other words, many forces resist outright repeal of monopoly protection. Present monopolists pay lip service to competition, but jockey for continuation of the status quo. They craft elaborate arguments that involve safety, reliability, and any other possible appeals to delay the process of opening competition. Their own self-interest is served by shaping the emerging competition to favor their companies. Many communities have municipal electric monopolies, or are served by Rural Electric Agencies, or by government-owned monopolies like the Tennessee Valley Authority. All resist competition. These realities make a quick move to free energy markets unlikely.

Many environmentalists fear that a free market for electricity will result in more generation at the oldest and dirtiest plants. A dirty old plant is also grossly inefficient. A free market will build new and better power plants that are cheaper to operate because they burn less fuel. These plants will be much cleaner because they burn less fuel and employ the latest pollution-control technology. The present approach to environmental regulation is a major barrier to efficiency. There are ways to regulate pollution that reward efficiency, and we should re-regulate in this manner. But change comes slowly.

Given the need to mitigate climate change and acid-rain pollution, and the political difficulty of quickly opening all electric generation, transmission, distribution, and sale to market forces, we have looked for a compromise solution. The proposed solution would work to promote cleaner technologies, would encourage the opening of markets, and would provide opportunities for all fuels and technologies. The proposed

Fossil Fuel Efficiency Standard would guide energy professionals of the world to a more sustainable and efficient future. It should be politically feasible to enact a single, nationwide Fossil Fuel Efficiency Standard that forces every firm in the electric-generation business to gradually improve fossil efficiency and to reduce dependence on fossil fuel. This proposed standard could be adopted by any country, and would soon cause that country to have an advantage in the energy costs of all of its products versus products from countries that continue to give monopoly protection to inefficient power generation.

Adopting a Fossil Fuel Efficiency Standard will lower carbon dioxide emissions and improve U.S. competitiveness versus other countries that do nothing. This standard and other suggestions herein can be adopted by any country that wants to increase its standard of living and improve its environment. In the process, countries will reduce their emissions of greenhouse gases and other pollution. Eliminating barriers to efficiency, ending all energy subsidies, and opening energy markets to competition will reduce electric costs by as much as 40 percent, will increase standards of living, and will move the world toward a sustainable energy future. Admittedly the environmental impact of a 45 percent efficient coal-fired plant is greater than that of a 45 percent efficient gas-fired plant. But the standard represents a compromise that perhaps can be embraced when more polarized approaches result in inaction. It is my hope that we will move forward and make balanced trade-offs between environment and economy that can be embraced by the powers that be. The alternative is a fight that the economy and the environment will both lose, even if we know we could do better.

The growing debate on mitigating global climate change is colored by the negotiator's choice of action. Representatives of most of the world's nations met in Kyoto, Japan, in December 1997 and adopted a protocol under which industrial nations would, if the treaty is ratified, agree to reduce their emissions of greenhouse gases. Since the major contributor to greenhouse warming is carbon dioxide, the conference framed discussions in terms of a carbon cap—a ceiling on the amount of carbon each nation would emit. A carbon cap sounds like invasive government regulation that we would very much like to avoid. Most people fear that this will lead to a "command-and-control" approach to energy, replete with more regulations and inefficiencies. The specter of intrusive regulatory mechanisms to enforce a carbon cap has created much opposition.

People react to this approach as though they have been told they have a terminal illness—they want a second opinion. They want to argue about the problem and delay the apparently painful cure until there is more certainty of the disease.

The carbon-cap approach focuses everyone on the science argument (Do we really need to reduce carbon dioxide emissions?) and on the cost of carbon reduction. This focus is unfortunate, because it makes ratification of the Kyoto treaty politically difficult and delays or kills the rational actions we need to pursue to end our fossil fuel waste. It is not clear to many people that we can both reduce carbon emissions and save money by increasing efficiency. They interpret global climate-mitigation efforts as expensive pollution control.

We have a better way. Convert the carbon cap into an efficiency standard, and the political debate changes. Forget fixed allowances of carbon and elaborate regulatory mechanisms. Simply require every generator of electricity either to meet nationwide standards of fossil fuel use per unit of useful energy produced or to purchase national credits from others who use less than the allowed fossil fuel. Then let the market decide hour by hour and day by day which fuels to use, which technologies to deploy, and how to achieve a scheduled increase in energy produced per unit of fossil fuel consumed. It is hard to argue against efficiency, especially when the proposed Fossil Fuel Efficiency Standard applies to every generator equally. An efficiency standard can work even if we do not open all sectors of electricity to competition, so it avoids many arguments. However, the standard will force all who generate power to adopt the efficiency strategies that they would have chosen under competition. It will force every power producer to confront every barrier to efficiency—and those archaic barriers will buckle under the collective pressure of the marketplace.

Some people argue that the government should not set new standards, but should just let the market decide which fuel to use, which technology to deploy, how much energy to use. We sit atop a mountain of old rules—monopoly rules, environmental rules, security-and-exchange rules, tax rules, and federal reimbursement rules. In a country that prizes efficiency, the Fossil Fuel Efficiency Standard provides a simple, elegant way to excise these many rules and to guide the United States toward a more sustainable energy future.

Humankind can enjoy a decent, sustainable standard of living without burning fossil fuel, if we allow enough time before that happens. Energy production from renewable energy, like wind, solar, geothermal, and biomass, is moving closer to economic reality. The sun gives earth 30,000 quadrillion British thermal units (quads) of energy every year. The earth produces about 3000 quads of biomass energy every year, whereas the total present global consumption of energy by humans is only 312 quads per year, or just 10 percent of the annual production of biomass. Clearly, we are not running out of energy. What we are running out of is room in the atmosphere to store the liberated carbon dioxide from fossil fuel.

Economists have built models that predict that all actions to mitigate climate change will cost the economy. But are they correct? The projections of economic harm from doing the right thing remind me of the response of a Boston federalist newspaper to Thomas Jefferson's Louisiana Purchase. It called Louisiana "a great waste, a wilderness unpeopled with any beings except wolves and wandering Indians. We are to give money of which we have too little for land of which we already have too much."[1]

Today's economic models often embody a similar lack of vision. The present inefficiency costs the economy. The economic models don't acknowledge savings from eliminating barriers to efficiency. They ignore many advantages of lessening our wasteful conversion of fuel to electricity and heat.

The vital step is to find ways to increase efficiency that represent good investments. Many energy entrepreneurs are prepared to make energy-saving investments.

One of those companies is Trigen Energy Corporation (New York Stock Exchange symbol TGN), of which I am a founder and Chief Executive Officer. We have a mission of producing heat, cooling, and power with one-half the fossil fuel and one-half, or less, of the pollution associated with conventional generation. Much of the logic and experience in this book comes from Trigen and its predecessor companies (Cummins Cogeneration Company and Cogeneration Development Corporation) that I have directed as CEO for the past twenty-one years. We find opportunities to combine the generation of heat and power, to automate boilers to increase their efficiency, or to recover and sell heat that was being wasted; we sign long-term contracts to supply cheaper energy. Our goal has always been to earn a profit by investing in energy efficiency. We have always had to provide savings to energy users to win their business. We think energy efficiency pays, and our record proves the point. Given a change, energy professionals will invest $200 to $300 billion in the United States economy to increase the efficiency of fuel conversion to heat and power.

There has been a wonderful experiment run over the past eight years, involving fifty-five million people, which proves the point. The United Kingdom under Margaret Thatcher began to deregulate electricity generation and sale in late 1989. Six years later, carbon dioxide emissions from U.K. power generation had fallen 39 percent, nitrous oxides had fallen 51 percent, and the price of electricity to all classes of consumers had fallen 15 to 20 percent. These trends are continuing. Similar proof that free markets reduce energy waste, reduce pollution, and save money can be found in the experience of Argentina and Chile where electricity has been deregulated.

Markets are wonderful when allowed to function. They constantly feed price signals and other information to all the participants, each of whom is free to act as he or she sees fit. Many actions of entrepreneurs prove to be foolish, or wrong, but successful actions are rewarded and rapidly emulated. The whole process is constantly self-correcting, unlike legislated solutions and central planning, which are static and inevitably based on yesterday's faulty projections. Market-based outcomes are constantly being improved by the action of collective, self-corrective intelligence, informed by prices. In fact, the economic health of every country in the world seems to depend on how free that country's markets are to perform their magic. We know this and, in general, embrace free markets. We know that central planning and regulation failed the entire communist world. Yet, we have hung on to centrally planned and regulated electric generation, which has also failed.

ORGANIZATION OF THE BOOK

This book opens with a general background of the climate problem and a discussion of the importance of energy to our well-being. In chapter 2, monopoly is examined with emphasis on why, between 1907 and 1922, all governments chose to grant exclusive monopolies to the electric industry. Monopoly regulation of electric production and sales has become, after ninety years, an unexamined paradigm in the United States, a perceived natural law. Several other countries have already ended their electric utility monopoly protection, and the United States is moving in that direction. However, monopolies are strongly entrenched in the United States, and our political system with its effective industry lobbyists and political contributions makes change difficult to achieve.

Chapter 3 examines the many powers humankind has already achieved over nature, powers that were formerly considered the exclusive domain of the gods. I argue that this knowledge carries with it an obligation to exert God-like intelligence over our actions. It is possible to develop God-like intelligence by connecting all human minds in a massive parallel processor of information. The incredibly complex process that is needed was invented thirty thousand years ago. It is called a *market.*

Chapter 4 focuses on the reasons that consumers make excessive use of energy and suggests ways that consumers could be educated and induced to change their behavior.

Chapter 5 shows how deregulation and carbon dioxide reduction are linked. The focus is on understanding how to change the behavior of the energy conversion industry. We first explain the basics of power generation, and then explain why combining heat and power production is absolutely essential to increase efficiency from any technology that involves burning any fuel.

Chapter 6 goes back to the global problem and deepens the explanation of what must be done to leave a decent legacy to our children and grandchildren.

Chapter 7 focuses on how to change consumer attitudes toward energy waste and energy use. It suggests a series of programs that would not cost much, but would make consumers aware of the value of reducing energy waste, enabling them to increase their standard of living while lowering pollution from energy conversion.

Chapter 8 spells out the barriers to efficiency that have been erected in the past ninety years. No one intended to create any of these many barriers, but rules and regulations that were promulgated to achieve other non-energy objectives are full of barriers to efficiency. Each barrier to efficiency is explained with some examples; suggestions are offered for modified regulatory approaches that could achieve the original goals and encourage fuel-conversion efficiency. Simply removing all of the barriers to efficiency will unleash market forces, trigger a great deal of improvement in efficiency, and change the behavior of the energy-conversion industry.

Chapter 9 considers the three federal actions that are currently being discussed to induce more fossil efficiency and suggests a fourth approach. What is being considered is a cap on carbon emissions with trading of permits, higher energy taxes, and a renewable energy portfolio. As discussed above, I suggest a fourth approach—a Fossil Fuel Efficiency Standard.

Chapter 10 proposes passage of ERRATA or the Energy Regulatory Reform and Tax Act, which includes all of the proposals herein to unleash market forces, and barriers to efficiency, and thus save money and carbon dioxide.

Throughout the book, I draw liberally on the experiences of people I have been privileged to work with in a quest to cut fuel use in half and cut pollution from power generation by more than half. Twenty years as an alternative energy provider—alternative to the monopolists—have brought me in contact with a fascinating range of courageous customers, visionary government leaders, and energy entrepreneurs who have dared to innovate to save energy and to reduce pollution and energy costs. Without naming individuals, I wish to thank them all by dedicating this book to those who are willing to try new and better ways to meet their energy needs. What this planet needs is more of these kinds of people who care about the human impact on the environment and are willing to act accordingly.

Special thanks are due to many people who have discussed ideas, read drafts, challenged conclusions, and encouraged me to put these thoughts on paper. Thomas Jensen, an attorney with Troutman Sanders and former White House Council on Environmental Quality Associate Director, has steered our thoughts and efforts and helped me understand the positive role of government. Mike Weiser, a Trigen vice president, has been a partner and part of the effort to change the way the world makes power for all of the past twenty-one years and has never failed to offer thoughtful and insightful advice and counsel. General Counsel Eugene Murphy has freely offered endless guidance in the law and few will ever appreciate his creativity in jumping the barriers to efficiency and persuading customers to enter long-term contracts. Mark Hall, Trigen's director of government affairs, was enjoying his responsibilities as manager of environmental health and safety when he was tapped to help craft and deliver our message to the Clinton administration in response to the president's call for ways to mitigate climate change without disrupting the economy. In the thousands of hours we have spent together answering this call, Mark has helped shape the arguments and test them with a wide variety of interested parties. He has chased data, challenged ideas, and nearly coauthored this book. Helen Hall has no special energy expertise but as could be guessed from knowing her son Mark, she is a thoughtful and concerned citizen whose comments on the entire manuscript were ever so valuable. George Keane, president emeritus of the Common Fund, first encouraged me as a fellow member of the executive committee, and later became Trigen's nonexecutive chairman. He has encouraged Trigen's actions and the writing of this book, offering innumerable thoughtful edits.

Sara McKinley is that special kind of journalist who reduces complicated energy issues to lay language. Her many suggestions have improved the arguments and language. Linda Prime has managed the process, chasing sources, keyboarding changes, challenging sentences, and still keeping a busy office functioning. Matt Steinwald, Mike Scorrano, and Rich Shaw have tirelessly unearthed data, which with skill and intellect they crafted and recrafted into explanatory visual graphs and charts. Finally, my deepest thanks to Judy, my wife of thirty-three years, who lifts me when I am down, trims my sails when they are too full, and has offered compassionate criticism and unfailing support each step of the way.

I hope the lessons and experience herein will offer guidance to policy makers and regulators about how they can lead our society to a sustainable future.

Thomas R. Casten
August 31, 1998

1

Establishing the Problem and a Market Perspective

GLOBAL-WARMING CONCERNS

In 1896, a Swede, Svante Arrhenius, was the first to propose a scientific theory that carbon dioxide emitted into the atmosphere from burning fossil fuels would act as a greenhouse gas. The sun's high-frequency incoming radiation passes through the carbon dioxide in the atmosphere and warms the ground, but the lower-frequency infrared radiation from the warmed ground bounces off molecules of carbon dioxide in the atmosphere, keeping some heat from radiating back into space. More atmospheric CO_2 could increase the global temperature.

Figure 1 shows world energy-sector emissions of carbon dioxide over the past 130 years. One can see the beginning of the Industrial Revolution, and then the major increases after 1940. There are temperature measurements for the same years, and after correcting for city growth (which raise local temperatures due to the "urban heat affect"), scientists have found temperature increases consistent with climate models.

By 1970, modest concern was developing about the increasing concentration of CO_2, but the climate models were not considered very good, and the global temperatures through the 1970s seemed to fall short of predictions. Then, in the 1980s, there were several very warm

Fig. 1. World Carbon Dioxide Emissions from Fossil Fuels[2]

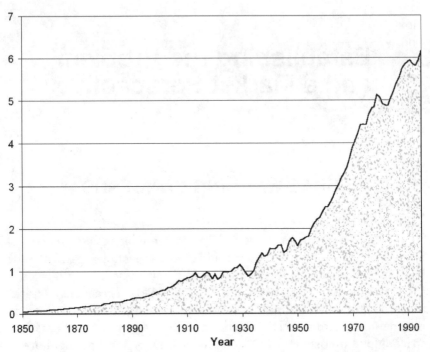

Billion Metric
Tons
of Carbon

years. Carbon dioxide emissions grew every year, as did concerns about global warming. In 1992, the first United Nations Climate Conference was held in Rio de Janeiro, Brazil. Despite much debate, a scientific consensus did not emerge that human actions were definitely impacting climate. Too many questions remained. A protocol was adopted calling for voluntary reductions of carbon dioxide to 1990 levels. The United States signed, but did not enact any meaningful programs to curtail fuel waste. Emissions grew. By 1997, the U.S. carbon dioxide emissions were 114 percent of 1990 levels and climbing.

Much has since been learned by the scientists. Computer capability has increased dramatically. Models have grown more sophisticated and have begun to account for other factors, including the negative impacts on temperature of some of human actions. Sulfur aerosols reflect some sunlight back into space before it can warm the earth, and this effect was factored into recent models. These models' fit to observed data has improved dramatically. These aerosols from burning coal temporarily offset the impact of increased greenhouse gases. The cooling impact of a ton of these aerosols vastly outweighs the warming impact of carbon dioxide in the short term, but the aerosols leave the atmosphere within one to four years, while carbon dioxide remains for hundreds to thousands of years. The improved climate models predicted that early fuel consumption increases during the Industrial Revolution would, on balance, not lead to warming due to the cooling effect of the added sulfur aerosols. The models went on to predict that by the late 1970s the weakly warming but lingering carbon dioxide would begin to overwhelm the strongly cooling but fleeting aerosols, and global temperatures would rise. This is exactly what has happened to global temperatures, with the warmest ten years of the past 140 occurring since 1980.

In 1995, a United Nations panel of twenty-five hundred scientists reached a consensus about the impact of burning fossil fuels and published a report that concluded, "the balance of evidence suggests a discernable human influence on global climate."[3] The debate is far from over, although the vast majority of climate scientists agree there is human-induced global warming underway. These scientists are probing historical data for better understanding and working to predict the possible impacts on global weather, but largely agree there is a problem. A few people with scientific credentials are climate-change skeptics, and many people who do not work in the field of climate express vociferous opposition to the climate theories. However, it is hard to tell whether those who question or ridicule the prevailing scientific view do so out of scientifically based doubt, out of concern for the presumed economic cost of lowering greenhouse gas emissions, or out of sheer terror.

The theory of climate change predicts that the warming will lead to weather extremes, colder and hotter, and will lead to more violent storms. Many scientists now assert that earth is experiencing these increasing extreme temperatures and violent storms. The El Niño effect during the winter of 1997–1998 has certainly disturbed many ordinary citizens in the United States and Mexico, and has brought home forcefully how much damage can result from climate change. The National Oceanic and Atmospheric Agency (NOAA) recently issued a special report on extreme winter weather and climate experience in the United States during this year's El Niño season, stating:

> This winter's El Niño ranks as one of the major climatic events of the century. The country as a whole saw the warmest and wettest January and February in the past 104 years. Rainfall records dating back to 1884 were broken in California, and temperature records from 1890 were broken in Ohio.
>
> When you look more closely at the numbers, you also see that this record-breaking El Niño is consistent with a worldwide trend over the last 40 years toward a warmer and wetter world. We can't draw a causal link between El Niño and global warming, but our modeling tells us that global warming may first manifest itself in changes in weather patterns. In other words, this winter's El Niño is a taste of what we might expect if the earth warms as we now project.[4]

During the summer of 1998, people had an unpleasant taste of warming. Forests burned in Florida, Texans suffered with day after day of over 100 degree Fahrenheit heat, and floods ravaged China.

CLIMATE-CHANGE HISTORY

In the early 1980s, a multinational project was established to collect and analyze samples of Antarctic ice that originally fell as snow up to 160,000 years ago. The objective was to obtain information on past climate and on how trace gases, notably carbon dioxide (CO_2), might affect climate. The end result was the Vostock ice-core data (so named because it was collected at the Russian's Antarctic Vostock base). The ice-core data provides a simultaneous record of climate and atmospheric CO_2 over 160,000 years, covering all of the last ice age and part of the preceding ice age.[5]

The two graphs in figure 2 chart the CO_2 concentration data and temperature data for the period from 160,000 years (on the right) to 300 years ago. The Vostock data stops at 300 years ago, so we have added the recent CO_2 data. During this period before 1950, the highest observed CO_2 levels were just under 300 parts per million (about 136,000 years ago) and fell to a low of 175 parts per million about 42,000 years ago. A rise to 275 parts per million over the next 40,000 years

Fig. 2. Atmospheric Carbon Dioxide Concentration
and Associated Temperature Changes[6]

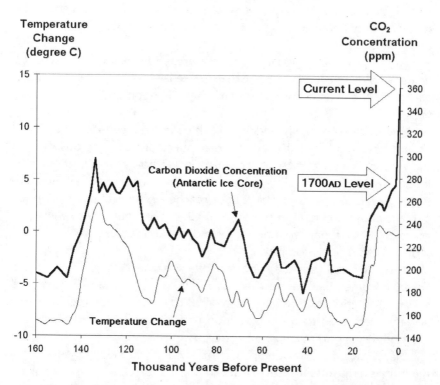

increased at a geologically slow average of only 2.5 parts per million per one thousand years.

The lighter line shows deviations in the same period from today's temperature in degrees Centigrade, and shows that most of the period was colder than today, by as much as 9 degrees Centigrade (16 degrees Fahrenheit). It is apparent to the eye that the temperature moves in very close synchronization with the CO_2 concentration.

A number of people have looked at this 160,000-year data to find patterns. One reviewer of the data did a mathematical correlation of the data and found the temperature moving up and down with CO_2 concentrations at a rate ten times as great as would be expected from the heat-trapping effect of CO_2 alone.[7] The speculation was that there are a number of indirect effects such as changes in cloud cover, changes in planetary reflectivity or *albedo,* and particularly an amplifying effect involving changes to atmospheric water-vapor levels. There is a great deal of uncertainty about how much influence these indirect effects will have on climate changes.

At the peak of the last ice age the average temperature was about 9 degrees colder than it is now. At that time a layer of ice extended all the way from the North Pole down to where London and New York are today.

Scientists do not fully understand the complex feedback effects and do not have models that predict the temperature changes observed in the Vostock ice-core data, so we must decide how to act without perfect knowledge. What is being learned is not comforting. CO_2 concentration levels in the atmosphere have reached 360 parts per million, 20 percent above any time in the past 160,000 years, and are expected to go to 540 parts per million in fifty more years, unless we lower the growth of CO_2 emission rates.

A small, self-interested minority argues that increasing CO_2 is good. Fredrick Palmer, CEO of the Western Fuels Association which mines and sells coal to utilities, used Earth Day 1998 to launch the "Greening Earth Society" dedicated to the proposition that CO_2 in the atmosphere will be good, leading to more abundant crops.[8]

Maybe not. Six days before Earth Day and Mr. Palmer's announcement, Massachusetts Institute of Technology researcher Maureen Raymo reported results of sediment cores recovered from the deep-ocean floor. Raymo and colleagues from Woods Hole Oceanographic Institute reported finding multiple occurrences of temperature swings of as much as 10 degrees Centigrade in just a few decades, back as far as 800,000 years ago.[9] Scientists don't know precisely what set of conditions triggered these sudden changes, but their discovery is sobering. We are traveling at high speeds and accelerating (increasing CO_2 emissions) on a road that we do not know. We may suddenly come to a cliff of rapid temperature change. We should also understand the size of the bet we are making. By not reducing CO_2, we are betting nothing less than the ability of the earth to support human life.

Fig. 3. World Population Growth[10]

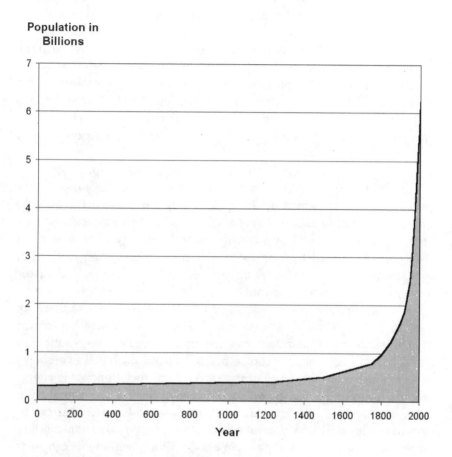

Population in
Billions

POPULATION GROWTH

For most of the time humans have been in existence the forces of nature were pretty evenly matched against our human desire to reproduce. Then, beginning four hundred years ago, the development of the scientific method enabled humankind to start gaining mastery over nature. Our advantage over competing life forms began between 1400 and 1600 C.E. and population began to grow.

Figure 3 shows just how incredible the population increase has been.

You see world population barely changing for sixteen hundred years. Famine, pestilence, disease, and other forces of nature kept human population in check. Then, as accumulating knowledge began to provide tools and strategies to enable humans to outwit nature, the world population soared.

The few people alive today who are over 100 years old have seen the world's population quadruple during their lifetimes. My mother was born in 1919, when world population was under two billion. As I write this manuscript, she is seventy-eight, and world population has grown beyond six billion. Geometric population-growth rates are very powerful. If Mom lives another ten to fifteen years, she may see world population increase by fourfold in her lifetime.

This growth of population and rising standards of living have increased energy use dramatically. *The use of energy to support our growing world standard of living is not synonymous with burning ever more fossil fuel.* Useful energy can be produced efficiently, or inefficiently. At present, we waste two-thirds of the energy in the fuel that is used to make electricity worldwide. We can increase energy efficiency and burn less fuel, providing we increase efficiency of conversion. Wasted energy does not increase anyone's standard of living; it simply pollutes.

It is also important to understand that the energy we use does not have to be produced with fossil fuel. There has been steady progress toward economically efficient capture of renewable energy, and our grandchildren may have no choice but to depend on renewable energy because we are burning up all of the stored fossil fuel. Renewable energy includes sources of energy which are replenished by the sun such as wind, hydro power, photovoltaic electricity, and all biomass that can be burned or digested into alcohol or other fuel. Our generation can help move toward economical renewable energy. Our governments can guide us toward a sustainable future.

GOVERNMENT ACTIONS

On October 22, 1997, President William Clinton gave a speech at the National Geographic Society Headquarters in Washington, D.C., outlining his proposed policies on climate change. He said that the science was clear that global warming is a problem, and that we must take steps to mitigate global climate change.

The president's call for action has fueled a raging debate, replete with multimillion-dollar advertising campaigns against doing anything, funded by companies that believe any steps to reduce greenhouse gas emissions will disrupt the U.S. economy in general, and their businesses in particular. The theoretical economists have weighed in with the supposed wisdom of their macroeconomic models of the economy. The models all assume that the market for energy is operating with economic perfection and that all useful investments to produce energy more efficiently have been made, given today's technology and today's prices for fuel and electricity. These macroeconomic models assume that society will burn less fuel only if governments increase taxes on fuel or mandate expensive changes in the way energy is produced and used—changes that the market would not choose without new government regulations. These theoretical models are dead wrong in their view that the energy market is operating with perfection and thus they try to answer the wrong question. The models explore the cost of various government actions such as taxes and "command and control" regulations that will mitigate climate change. Economists seek to identify the least-cost actions. The models need to be rebuilt to explore savings resulting from eliminating barriers to efficiency.

As mentioned in the preface, many people look at these projections of economic pain and act like they have just been told they have a terminal illness—they want a second opinion. They want to quibble about the science of global warming. Demonstrating that reducing CO_2 emissions will save money and reduce pollutants casts the global-warming scientific debate in a completely different light. ***We should improve energy efficiency for business reasons—regardless of our views on global warming.***

What follows is an explanation of why the United States has failed to adopt optimal strategies for efficient energy production or for investing in energy efficiency. Wherever possible, the discussion is rooted in personal experience of trying to improve efficiency—trying to change the way we make and use power. The actions recommended here to reduce our use of fossil fuel and our waste of energy make economic sense and will improve the standard of living of everyone. The inevitable by-product will be to mitigate climate change and to stop killing the trees. These actions will not disrupt the economy, and they are not simply a "least regrets" strategy for addressing climate change. They will improve both the economy and the atmosphere.

Climate scientists have raised a very credible but nonetheless debated alarm. It has forced us to examine whether we need to continue depleting fossil fuel reserves and increasing CO_2 concentrations. This examination has revealed many artificial barriers to efficiency; we are paying a high price for waste.

2

Understanding
Monopoly Behavior

WHAT IS "MONOPOLY"?

We all know the word *monopoly* and have some idea of what it means, but probably lack detailed understanding of how business pursues its profit goals under monopoly protection. Monopoly protection of the worldwide electric industry has led to extremely low efficiency, unnecessary pollution, and most importantly, to unnecessary carbon dioxide emissions. I have had the somewhat unique experience of trying to introduce efficiency and lower the costs of electric generation for the past twenty years. My colleagues and I have been constantly frustrated by a plethora of rules, regulations, abuses of power, and other barriers to competition erected by well-meaning government officials within a system of monopoly protection. Understanding monopoly behavior is essential to understanding how the system must change.

Parker Brothers' ever-popular board game, Monopoly, teaches us the most important lesson about monopoly. The winning strategy is to acquire all of the properties of a color group or industry, such as all the railroads or all the utilities. When you own one railroad, the rent is $25. When you acquire a second railroad, you are beginning to control the market and the rent goes up by 100 percent to $50. Own all four railroads, and you monopolize the railroad market, so you charge a

25

"monopoly rent" of $200! This reflects economic reality at a simple level. The rents are modest when there is competition, but as the industry concentrates in one owner's hands, prices rise. Consumers pay because other choices, like walking or hauling freight with a horse-drawn cart, are more expensive. The monopolist owner of all the railroads loves this situation, because he or she can charge prices just below the next best alternative, and earn a large profit. In time, others see the huge prices and develop alternative forms of service, but this takes time. New services and ways to meet consumer demands will always emerge in time. When government steps in and declares the existing service an official monopoly (and uses police power to prevent competition), the process of developing new approaches and new services is slowed. A growing body of rules shuts out alternate would-be providers. Such is the history of all regulated economies.

In the nineteenth century, many new businesses grew and combined, hoping to gain market domination and thus be able to charge monopoly rents. Everyone is aware of the success of John D. Rockefeller in gaining control of oil production, and of Jay Gould, J. P. Morgan, Andrew Carnegie, and others in gaining significant shares of major industries like steel making or banking. Journalists coined the term *Robber Barons* and delighted in depicting the rapacious behavior of these entrepreneurs. The Sherman Antitrust Act was passed by Congress in 1890 to limit the behaviors these would-be monopolists often employed to gain market dominance. The original law was deemed too weak, and in 1914 Congress passed the Clayton Antitrust Act, which toughened government's powers against restraint of trade. Antitrust rules have been applied to almost all industries ever since. As I write this book, the newspapers are filled with Justice Department actions against computer software giant Microsoft, the corporation that developed the immensely popular "Windows" software to facilitate the operation of personal computers. Microsoft's alleged violation of antitrust rules was to demand that computer makers simultaneously install the Microsoft Internet browser with the Windows operating system. This is termed *bundling* under the antitrust laws, and a court has ordered Microsoft to stop the practice. I make no comment on the merits of the case, but use it as an example of how unregulated industries face limits in what they can do to exclude competitors.

Another example further illustrates the antitrust prohibitions against bundling products. Eastman Kodak found that independent service and repair companies offered their services for as little as half of what Kodak was charging. To protect their service and repair business, they refused to sell repair parts to the independent companies.[11]

The U.S. Supreme Court struck down the practice as a violation of antitrust rules. In essence, it said this bundling prevented other suppliers of repair services from having a fair chance at entering the business and was illegal. These limits to what any supplier of any good or service can do to block competition apply to all businesses except government-created monopolies.

Some monopolies remain, to this day, quite logical to everyone. It is hard to imagine how we would benefit from two water mains in the street, or two sets of sewers. Other historically "logical" monopolies of the Sherman Act–era have been swept away by technology. Rail was a vastly better way to travel and to ship goods when the only alternative was a horse-drawn carriage on a dirt road. Competing rail lines seemed inefficient. The Interstate Commerce Commission kept railroad monopoly rules in place until 1970, long after the diesel-engine-powered forty-ton trucks took advantage of the new interstate highway system to capture most freight business. Technological advances make the law obsolete, leaving consumers to suffer until laws are (sometimes) repealed.

In 1978 after fifty-six to ninety years of monopolistic electric generation, transmission, distribution, and retail sale, the United States Congress passed the Public Utility Regulatory Policy Act (PURPA). This law allowed certain non-utility generation, but kept all other aspects of electricity a monopoly. In 1989 Margaret Thatcher persuaded the British government to allow competition in electric generation and retail sales, but not distribution. In 1991 Argentina opened its full electric sector to competition. Then Sweden, Norway, and Chile opened their electric markets to some competition. These first few partial deregulation actions came decades later than warranted by new technology, but they already provide proof that when given a chance, market forces lower costs, fuel use, and pollution.

Most people believe the antitrust statutes are essential and ensure a level playing field to prevent incumbent firms from abusing their market power to limit new entrants. However, when there is a government-created and protected monopoly, such as the provision of electricity, antitrust rules generally have not been applied. For example, most electric utilities offer a bargain "all-electric" rate if the building owner agrees to use electricity for heat, cooling, and lighting—for all possible uses. The oil dealer and the gas company offer competitive ways to produce the heat and cooling for the building, but are prohibited by monopoly law from selling the electricity that the building must have for lights and motors. Fuel dealers would like the building to install its own boilers and use gas or oil to make heat. One cooling technology uses a gas-fired absorption technology to cool the water, and gas distribution companies promote this solution in order to increase the sales of gas in the summer. District energy companies build central boiler plants and/or chilled water plants and produce steam, hot water, and chilled water which they send through buried pipes to various buildings, eliminating the need for separate energy plants in each building. These district energy companies would like to heat or cool the buildings, but are also prevented from supplying electricity.

These three types of companies offer a competitive service. The basic notion of the antitrust laws is to prohibit one provider from using monopoly power to block the offerings of a competitor. Utilities usually cite the costs associated with their "obligation to serve" as justification for holding customers hostage. But they continued until recently to build new power-generation capacity even though many customers wanted off the system.

Worse, some states still regulate steam and hot-water distribution. Even though these services are clearly not a monopoly, they are subjected to burdensome regulation and denied the right to sell electricity to their customers. The playing field is *not* level. Alternate providers of energy are prohibited from selling electricity to their users so they are prevented from making innovations that combine electricity with heating and cooling supply. As a result, the United States and other countries make inefficient use of fuel and produce needless pollution, especially carbon dioxide.

Much of what is wrong with our energy supply system (and most of its fantastic waste of fossil fuel) is a result of government actions. As strange as it seems, many opportunities to save money are blocked by well-intentioned laws and regulations. We list and explain many of these unintended barriers to energy efficiency in chapter 8. First, we need to understand why governments all over the world decided to enact monopoly protection of electric generation and distribution.

Economists have traditionally identified certain business activities as *Natural Monopolies*. These are businesses that, from the vantage point of the economists at the time, contain unique features that make it inefficient to have more than one supplier. We emphasize "at the time" and "from the vantage point of the economists." Technology development and deployment, driven by profit-seeking entrepreneurs, have a rich history of bypassing these natural monopolies and introducing competition.

An example is telephone service. From Alexander Graham Bell's invention of the telephone until twenty-five years ago, economists and government officials could visualize only one way to provide telephone service. This "unique" way was to connect each subscriber with a copper wire to a central network that in turn was connected to every other subscriber. It seemed that duplicating copper wires on every street would be most inefficient, and we deemed telephone service a natural monopoly. As far as I can tell, we were initially correct, based on the technology available at the time. The good news is that new technologies develop constantly. Inventors and developers are always seeking new ways to provide desired services. The bad news is that laws do not change automatically. They are often still in force for decades after new technology invalidates the original premises of the law. The telephone example demonstrates both points.

Entrepreneurs deployed new technology that bypassed the wires for long distance. They developed towers that transfer voice and data from point to point by using microwaves, thus competing with wires for long distance. In the early 1960s, satellites were placed in orbit that could receive communications from one spot on the globe and send the information to a receiving dish thousands of miles away. Corning and others have developed an entirely new transmission technology of glass fiber-optics that uses light waves to carry information. These fibers carry many times more information than copper wires and are the thickness of a human hair. Packed in bundles, they transfer vast amounts of information. A number of companies have found it economical to lay thousands of miles of this new glass fiber. Williams, a Tulsa, Oklahoma–based company, was a natural gas and petroleum products pipeline company which realized that its decommissioned pipes had added value as a place to install fiber optic cables. In 1986, Williams became a long-distance communications giant.

Many years after the emergence of these new and competitive ways to move information and voice communications over long distances, some governments opened the long-distance communications market to competition. *In ten years after the initial loosening of long-distance telephone regulation in the United States, the real prices of long distance telephone services fell by 47 percent.*[12]

However, these incredible benefits of long-distance telephone competition remain widely prohibited around the world. The benefits of telephone competition are denied or severely limited by governments out of a misguided strategy of protecting their native industries. The same government often owns the telephone company it is protecting, which further compromises decisions. These government-owned monopolies almost always charge higher prices and offer fewer options than is the case where competition has been allowed. If you doubt this, go to France or Japan and call the United States from your hotel room. Take lots of money or a credit card with a high credit limit for this experiment.

Political science professor Theodore J. Lowi at Cornell University wrote in *The End of Liberalism*[13] that government agencies become captives of the interest groups they serve over time. It is fascinating to review his analysis, which was developed for government in general, and apply that analysis to the specific case of regulatory agencies and the special-interest groups subject to their regulation. The special interests, in this case the regulated monopolies, capture the regulatory commissions. These businesses are the natural place for the commission staffers to go to work when they leave the regulatory commissions.

Both parties develop their own language and protocol, which is difficult for others to grasp or manipulate. As more and more precedents are established over time, the regulatory commission rarely questions past practices. As Lowi says, over time the regulated capture the regulators. Next, the agencies add rules and procedures that require more staff to supervise, and the new staff finds more "useful" ways that they can carry out their regulatory mandate. In other words, they extend their reach. The monopolies come to rely on this regulatory regime, which becomes bureaucratic and ossified.

After some years, or decades, new technology renders invalid the original premises that justified the monopoly protection, but this new technology holds little allure for the existing monopolies. Monopolists have a guaranteed market and it is not in their interest to adopt technology that undercuts the logic of the monopoly. Other firms would like to enter the market, and could use the new technology to bypass the monopolist. This new technology has the capability of providing consumers with many benefits, including price reduction for existing services, quality improvement of existing services, and new services. However, the old and often unchanging laws make it illegal for the new entrant even to offer the services. This process inexorably drives the regulated industry further and further from optimal. The new technology may not help the monopolist, who already enjoys 100 percent of the business. The laws prevent those firms who want to enter the business from competing. New and better technology often simply wastes on a shelf. A few special interests—the monopolists—are protected at the expense of all consumers. Indeed, the regulated industry begins to leave the ground littered with $20 bills, contrary to the macroeconomic model's picture of functioning markets.

Government-owned France Telecom makes a profit every year, and pumps out plenty of public-relations information to tell everyone how great they are, but French businesses and consumers are forced to pay noncompetitive prices for communications and in many cases remain stuck with analog service. Analog phones emit clicks equal to the number dialed, instead of special dial tones for each number. Do not call your firm's automated telephone-answering and voice-mail system from a phone in India, Nepal, or a whole host of other countries with government monopoly telephone services. You will not be able to send the digital tones a modern system needs to connect you to your own phone.

After years of denying competition, the United States has led the way to deregulate long-distance communications and opened the market to MCI, Sprint, ATT, and other long-distance providers. However, competition is prohibited for local service, and the local telephone service is even further behind available technology; available new technology is not allowed to compete in local service. Today, our cell phones don't need wires, our cable television connections are also capable of carrying voice and data communications, and it may soon be possible to carry voice and data communications on electric wires. Nonetheless, we continue to protect local telephone service as a natural monopoly. The result is high-priced local service and bloated local telephone company staffs. Although consumers are changing rapidly to a dependence on data communications (relying heavily on the Internet) the local phone companies have been very slow to improve their technology to provide the necessary bandwidth or information-carrying capacity. As the disparity between competitive possibilities and reality grows larger and larger, governments may finally open local telephone markets, allowing competitive market forces to determine the best way to serve our needs. If history is any guide to the future, this process of opening the markets will be delayed for years beyond the actual end of any natural monopoly.

A BRIEF HISTORY OF ELECTRIC MONOPOLY PROTECTION

Now that we understand how monopolies in general work, let's look at how these monopoly protections have been applied to electric distribution, transmission, and generation. Let's start with the basics—why electricity monopolies were created and why government agencies continue to think that they do a good job of controlling monopoly rents. (Economists use the term *Monopoly Rent* to describe the extra price a monopolist can charge for its goods or services, in the absence of competitive market forces.)

Electricity was not developed for commercial use until 1880. The electric industry evolved over the next thirty years as any other new industry develops, i.e., with many entrants. Some cities established government-owned systems—a tradition that continues today—as others opted to give exclusive service rights (franchises) to private companies. These "franchised" monopolies eventually submitted to regulation of prices and the conditions of service at the hands of state public utility commissions that were established in almost every state between 1907 and 1922.[14] Electric generation, distribution, and sale were made a protected monopoly by governments in the United States and later around the world. These services were not again subject to market forces anywhere in the world until PURPA opened some generation competition in the United States in 1978, and the 1989 Electricity Act began to deregulate United Kingdom electricity generation and supply. Over the past eight years Britain, Argentina, Sweden, Norway, and Chile have repealed monopoly laws and gradually allowed anyone who wished to supply electricity.

In the early part of this century, there were good reasons to set up monopoly protection for electric generation and distribution—reasons so good that every single country in the world established monopoly protection. At the turn of the century, the assumption was that the most economical way to generate electricity was in central plants and then transmit that electricity to consumers instead of relying on small generating plants in each customer's facility. If one accepts this assumption, then electricity distribution has a characteristic economists call a *natural monopoly*. We do not presently know of cost-effective ways to deliver electric power to a user from a distant generating station except by using copper or aluminum wires. Limited by our assumption that central plants are the most economical (or perhaps the only) way to produce electricity, it doesn't seem to make sense for society to run competing wires down each street so that the users can choose who delivers their electricity.[15] If instead, one assumes we could build local or "dispersed" electric generation that is more economical than our distant central electric-only generation, the idea that electric distribution is a natural monopoly becomes questionable.

In fact, my colleagues and I have built a business of installing and operating dispersed combined heat-and-power plants that offer both heat and power at prices lower than conventional supply. Many other independent power companies are daily proving that the basic assumption of monopoly protection of the electric industry is flawed. Small, mass-produced combined heat-and-power generating plants that use oil or natural gas offer a competitive way to provide electricity and heat. These plants use half the fuel that the central, electric-only generating stations use, and thus emit half as much greenhouse gas, i.e., only one-half as much carbon dioxide.[16] Curt Yeager, president of the Electric Power Research Institute, has predicted that in ten years, nearly all new electric generating plants will be dispersed and will be solar, wind, fuel cells, or combined heat-and-power plants. These generation technologies severely weaken the logic that electric distribution is a natural monopoly. However, opening electricity distribution to competition is tomorrow's work. Today, we have a more pressing problem.

Diffusion of Innovations Takes Time

The diffusion of innovations simply doesn't happen quickly. In Edison's day, potential users were not familiar with electricity. They had lived all their lives without electric power and were not ready to sign long-term contracts for a product they were not sure they needed. There is an aca-

Fig. 4. Diffusion of Innovations[17]

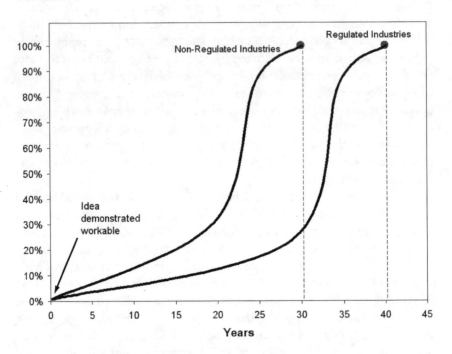

Market
Penetration

demic field of study that deals with the diffusion of innovations, and studies all the reasons why innovations take a generation or more to replace old, inferior approaches. The unambiguous findings of these studies are that a few people begin to test an innovation with a trial, and the innovation expands in use only if the trial results are positive. It is not uncommon for up to thirty years to elapse from the date of first commercial demonstration of an innovation to total adoption by society. There are many examples of this slow diffusion of innovations. Figure 4 portrays the typical adoption curve for new ideas to penetrate the marketplace for regulated and unregulated industries. Note that the adoption curve is about ten years longer for innovations to be fully accepted in regulated industries. For example, the aluminum grain car enabled railroads to haul more grain in each car than in a heavier steel car, and changing to aluminum cars made economic sense, but it took forty years after commercial demonstration before the regulated railroads had totally replaced the steel grain cars. With competition, diffusion of useful innovations is completed in thirty years or less.

In the late 1920s, hybrid seed corn was developed that significantly increased the yield per acre. Prior to the development of hybrid seed corn, farmers had always saved some of their crop for seed to plant the next year, but this practice does not work with hybrid seed. The kernels of corn from a hybrid crop do not reproduce true to the parent and are useless as seed. Hence, new seed must be purchased every year from the seed companies who specially produce the hybrid corn. It took an Iowa community thirteen years from the first introduction of the vastly superior hybrid seed before all of the farmers finally switched from saving their own seed. Until the time passed and all farmers had switched to hybrid seed, there were no guarantees that all farmers would switch. In fact, every farmer in the study first used hybrid seed on part of their acreage, then with success, expanded the use of the hybrid seed to more acres. This diffusion-of-innovation process is essential to understand if we are to learn why electric service was turned into a monopoly.

The time delays are rooted in human nature. The example of the transistor will help to illustrate this point. William Shockley, John Bardeen, and Walter Brattain invented the transistor in 1947.[18] Over the next twenty-five years the transistor slowly gained acceptance, and slowly replaced the larger, less efficient, failure-prone vacuum tube. When I was ready to leave Vietnam in 1967, twenty years after the transistor's invention and demonstrated commercial use, I went to the base PX and purchased a state-of-the-art Sansui stereo receiver and amplifier. It was filled with vacuum tubes. By 1970 tubes were gone from all new electronic products. Ten years later, tube testers disappeared from drug stores. (To see vacuum-tube computers still at work today, visit your local air traffic controller. It won't help your attitude toward flight safety).[19]

Power entrepreneurs knew intuitively about the painfully slow diffusion of innovation. They knew they would have to invest money to string wires for their new electric power down block after block, with few, if any, initial customers. In time, they hoped they could persuade the people whose houses and businesses their wires passed to switch from gaslights and candles to electric light bulbs, and thus increase electric sales. Keep in mind that the users could not just say, "Yes, I want electricity; please turn it on." First, these would-be electric customers had to invest money in an electric service panel, in new wires inside their buildings, and in new light fixtures and bulbs. There were no listings in the Yellow Pages for electricians or electrical contractors since they did not yet exist. The local hardware store did not stock all the electric paraphernalia that we take for granted today. Potential customers were not sure that electricity was safe, would be cheaper, or would be worth the bother. These consumers had functioned all their lives with candles, whale-oil lamps, and gaslights. Furthermore, and critical to the adoption of innovations, none of their friends or acquaintances was using electricity. Most people do not like to be pioneers.

A Whale (Oil) of an Example

I recently met the president of one of the oldest oil distribution companies in America, who had reviewed the fuel-oil-pricing records of the past 120 years. He found that there had been a sharp increase in the price of whale oil in the mid-nineteenth century. *Scientific American* wrote on June 27, 1857, "the whale oils which hitherto have been much relied on in this country to furnish light and yearly become more scarce, may in time almost entirely fail."[20] They went on to note that the increase in U.S. literacy was spurring a demand for lighting. Two years later, in 1859, an unemployed railroad conductor, "Colonel" Edwin Laurentine Drake, used salt-well drilling equipment to dig into oil-bearing strata in Titusville, Pennsylvania. His seventy-foot well struck oil on August 28 and was soon producing four hundred gallons per day.[21]

This bit of background illustrates two important points. First, the fact that it occurred 120 years ago, prior to electricity, helps us remember that electricity is a recent development and in part responded to a growing need for artificial lighting to supply an increasingly literate population. Second, it illustrates how markets work to fill unmet needs.

In fact, the story continues before electricity was developed. The next year, in 1860, a Cleveland produce firm sent their junior partner to investigate the potential of the petroleum found at Titusville. His official report was that petroleum had little future, but as is often the case, we must watch people's feet, not their lips. This young man pooled his savings with those of his partner, Maurice B. Clark, to invest $4,000 in the lard oil refinery of candle maker Samuel Andrews, with an eye to converting that plant to refine petroleum. He persuaded richer men to provide capital to build more refineries for petroleum, which he foresaw as a major energy source.[22] This young man was John Davidson Rockefeller, and the rest is history.

Electric light is a great substitute for whale-oil lamps, but first, the electric light bulb had to be invented. Thomas Alva Edison patented an incandescent lamp in 1879, and in 1880 began constructing the world's first central electric power plant in New York City.[23]

Problems of Early Electric Entrepreneurs

Consider the problems that Edison and other early electric-industry pioneers faced. The fractional-horsepower electric motor was not yet invented. Consumers did not yet have available, nor could they even imagine, all of the marvelous devices that small motors power, such as refrigerators, elevators, fans, and air conditioners. Only after electric distribution systems were developed and had reached reasonable numbers of buildings would other inventors find ways to use electricity to meet consumer needs. Entrepreneurs would, over time, develop many new applications. Radio was not yet invented, nor were televisions, copiers, or computers. Early acceptance of electricity was a very tough sell.

The *New York Herald Tribune* published an article on October 15, 1911, "Will Tesla's New Monarch of Machines Revolutionize the World?"

> "There was a time when men of science were skeptical—a time when they ridiculed the announcement of revolutionary discoveries.... Times have changed since then, and the "practical" engineer is not so incredulous about "scientific" discoveries. The change came about when young [Nikola] Tesla showed the way by which the power of Niagara Falls could be utilized. The right to divert a portion of the waters of Niagara had been granted; then arose the question of how best to utilize the tremendous power thus made available.... An international commission sat in London and listened to theories and practical plans for months.
>
> Up to that time the only means of utilizing electric power was the direct current motor, and direct current dynamos big enough to be of practical utility for such a gigantic power development were not feasible.[24]
>
> Then came the announcement of young Tesla's discovery of the

principle of the alternating current motor. Practical tests showed that it could be built—that it would work.

That discovery, at that opportune time, decided the commission. Electricity was determined upon as the means for the transmission of Niagara's power to industry and commerce. To-day a million horse power is developed on the brink of the great cataract, turning the wheels of Buffalo, Rochester, Syracuse and the intervening cities and villages operating close at hand in the great new electro-chemical industries that the existence of this immense source of power has made possible, while all around the world a thousand waterfalls are working in the service of mankind, sending the power of their "white coal" into remote and almost inaccessible corners of the globe, all because of Nikola Tesla's first great epoch making discovery."[25]

But stop and consider the commercial considerations for the developers of such "white coal." What were entrepreneurs, lenders, and investors to rely upon for revenue if they invested time and capital to divert Niagara Falls, build generators and transmission lines, and develop a new electric industry? Governments wanted to speed the deployment of these wonderful advances. But how?

Problem Solved—Exclusive Franchises to Sell Electricity

What these early power entrepreneurs needed was a franchise that granted them the sole right to sell electricity in the franchise area. With this exclusive franchise in hand, a power entrepreneur could find some banker willing to loan money, believing that in time, enough people in the franchise area would see the value of electricity and would purchase sufficient electric power to support repaying the loans. The exclusive franchise guaranteed that when people did adopt electricity, they would have only one choice of supplier—the firm that had borrowed the bank's money to establish the local electric infrastructure. Granting a monopoly franchise seemed like a good way for cities to encourage electric infrastructure development with private money. City governments saw granting electricity franchises as a way to spur private investment that would improve the quality of life in their city. I think the governments of the time were probably right: granting monopoly franchises all over the world greatly increased the diffusion of electricity as an innovation.

Controlling the Monopolist's Greed and Politician's Income

There was, however, a further complication. How would the governments keep the entrepreneurs with their exclusive franchises from overcharging—from ripping off the public they were supposed to be serving? Governments needed to appoint Plato's ideal "philosopher king," who would fairly grant and administer these franchises to ensure that society would be well served.

Plato notwithstanding, government is and will ever be grounded in the real world of human frailty, and not run by philosopher kings. In the real world, electric-industry entrepreneurs found they had to deal with city councils, aldermen, and mayors. The first approach to keeping the electric company honest was to grant an exclusive franchise to sell electricity only for a limited time span, often for only five years. Governments would have a chance to review the performance of each power company, and to renew electric franchises only where justified. This grant of five-year franchises guaranteed the need for relatively frequent "votes" for franchises. Some elected officials inevitably realized that without their vote or support for granting and continuing the monopoly franchise, the power entrepreneur would have no business. As Lord Acton said, "Power tends to corrupt, and absolute power corrupts absolutely."[26]

In the wide-open world of the early 1900s, some local aldermen found that they could augment their meager council incomes by selling their votes. When society does not, or cannot unleash market forces to control business, alternate institutions are needed. It takes time for a democracy to develop institutions that are graft-free and reasonably efficient in controlling human greed. I find no examples of any institution that ever worked as well as market forces to control greed and direct everyone's self-interest toward the common good. When we substitute government control for market competition, the first "solutions" to perceived problems often create more problems, and this was the case with the granting of short-term exclusive electric franchises. The constant renewal of local electric franchises was neither efficient nor free of graft. An alternate way to control the power entrepreneur would be to regulate the business prices, terms, and conditions of electric sale, but public utility commissions had yet to be invented.

To make matters more complicated, the city councils usually divided their town into ever-changing territories, and granted five-year franchises to different firms in different areas with different renewal times. As the electric industry began to consolidate; numerous small firms bought other firms, or merged, accumulating a variety of these franchise grants. They were forced to constantly engage in "purchasing" the votes of aldermen to keep their electric franchises intact. A better solution was needed, and would be proposed by one of the most successful electric-power entrepreneurs.

Samuel Insull Invents State-Granted Monopolies

Samuel Insull (1859–1933) emigrated to the United States from England in 1881, managed Thomas Edison's industrial holdings and eventually co-founded the company that would become General Electric. In 1892 Insull moved to Chicago where he began to assemble his empire of utility companies. He faced extremely frustrating local politics in Chicago, a city then known for its political chicanery. Insull literally invented a better way—at least a better way for his electric companies. He convinced the legislature of the State of Illinois that electricity was a natural monopoly, and that it was vital for society to become rapidly electrified. This created much larger and more permanent franchises and eliminated the constant drain of paying graft to greedy aldermen. His Chicago holdings eventually included Commonwealth Edison, Peoples Gas, and Northern Indiana Public Service Company, plus many more utilities.[27]

To grant a monopoly franchise, the legislature needed a new way to control the monopoly. The solution for Illinois legislators, and indeed for other states and countries that sought to regulate electricity, was to appoint a body to limit the profits that could be earned by the monopolist. The basic mechanism that emerged to supposedly protect the public's interest remains largely intact ninety years later. Government would appoint (in some states elect) a public service commission, which would regulate the electric utility monopoly and approve rates charged for power, decisions to add generation, and all major debt undertakings. These commissions would allow electricity prices high enough to cover all of the electric company's costs and provide a "fair rate of return" on the monopolist's invested capital. In theory, these public service commissions would be able to understand what were legitimate costs, what technology was available, and what other comparable businesses were able to earn on capital. In fact, as the world became more and more sophisticated, the understanding of these commissions has fallen further and further behind. Their decisions are often far from optimal, and never quick.

The mechanisms of a regulatory commission were intended to curb monopoly rents, and to encourage economies of scale and rapid electrification of the community. Most people place a lot of trust in this widespread approach to regulation. I am not among that number. Society's present separate generation of power and heat is far from optimal, and in fact fuel and cost efficiency is severely constrained by the very regulatory apparatus that was put in place to protect the consumer. The widespread commission bias against replacing old plants before they have worn out beyond repair has led to an aging fleet of inefficient and needlessly polluting power plants. It is time to move on to the best form of regulatory control ever invented—free-market competition.

It is not clear that competition could have worked well until thirty years ago. We did not have the technology or availability of convenient fuels to facilitate electricity competition before about 1965. We needed a substitute for competition in the early days of electricity, and regulation carried out by regulatory commissions was the usual answer in most countries.

Some countries decided to go further and ban the private supply of electricity. They created state-owned monopolies that were often given the further power of regulating themselves. This incredibly bad idea has actually been used in many parts of the world with predictably poor results. If there is any doubt about the wisdom of self-regulating state electric bodies, it is instructive to look at an example north of the United States border. Ontario Hydro has the monopoly right to generate power for the entire province of Ontario, Canada, and was further given, by law, the responsibility for regulating itself. Ontario Hydro managed to raise rates in the 1990s while the rest of North American suppliers were lowering rates, and in 1997 had to close down 4,600 megawatts of nuclear generating capacity because it was not safe or economical. The taxpayers in Ontario are paying for this mistaken approach to regulating electricity.

The United States has its own examples of regulatory failure. New York's Long Islanders have the distinction of being the worst hit, paying thirteen to sixteen cents per kilowatt-hour (kWh) for inefficiently generated power and debt repayment for the Shoreham nuclear plant, which never opened. These same Long Islanders may win the prize for consumers most ravaged by bad governance. The plan to have the state-funded Long Island Power Authority (LIPA) buy the local utilities' transmission and distribution system and lock in noncompetitive high rates has been hailed by Governor Pataki.[28] However, this plan will effectively eliminate the benefits of market competition for the next fifteen years.

But I am ahead of myself. The benefit side of establishing electric monopolies either by granting franchises to private entrepreneurs, or by establishing state- or city-owned electric companies was very clear in the early days. Electricity was a boon to society and it was very desirable to electrify all the homes and factories as fast as possible and extend electric service to as many as possible. Since the power business was and is capital intensive, the cost of capital, i.e., the interest rate on borrowed money and return on equity, strongly impacts the price of electricity. If consumers are to enjoy low-priced electricity, it is critical to limit the cost of capital used by limiting the risk of repayment. When a power company has a franchise that extends forever, investors and lenders see electricity as a low-risk business and accept lower returns or lower rates of interest than on loans or investments in other businesses that are more subject to competition. Taken alone, this single feature of electric monopolies (not having to pay high rates of return to investors and lenders) would lower the cost of power and it seemed a persuasive argument for regulated monopoly. (A slightly less than optimal fuel efficiency brought about by no competition quickly offsets the interest savings, but optimistic legislators and regulators also thought they could to ensure efficient production. They were wrong.) Granting franchises induced businesspeople to make large investments and more rapidly electrify an area, and this was clearly a societal good.

Economies of Scale

There were for years economies of scale that justified government's view that society would be best served by a monopoly utility. This would give the monopolist aggregated loads and let the utility build larger plants that would have lower costs. Technology changed all of this, and Charles Bayless, past chairman of UniSource (the parent company of Tucson Electric Power) and current chairman of Illinova Power, provides one of the best explanations. His own words follow figure 5, which is his graph of optimal plant size over time.

Fig. 5. Optimal Plant Size

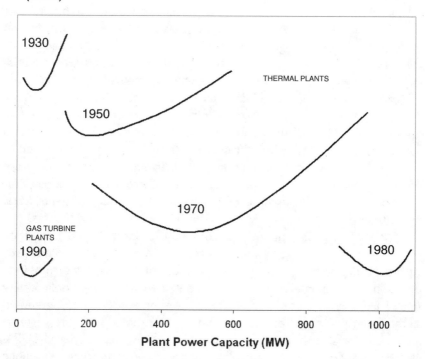

Figure 5 "... shows how the short-run average cost (SRAC) and size of installed capacity has changed over time. In 1930, a 50-MW unit was probably cheapest to build on a $/kW basis. Units sized below 50 MW would have imposed higher engineering and research and development costs. By the 1950s, technology had improved and the cheapest unit produced 200 MW, by the 1970s, 500 MW. This steady march of optimally sized units to larger capacity peaked in the early 1980s with 1,000+ MW units. But in mid-1980, a startling thing happened. The size of the cheapest plant dropped dramatically. Today the cheapest unit is a gas turbine in the 50- to 150 MW range ... the important point is that the optimum size has shifted from 500+ MW (10-year lead time) to smaller units (1-year lead time)."[29]

Given the perceived economy of scale, it seemed logical to governments and commissions to grant monopoly franchises for both the distribution and generation of power. Power entrepreneurs, eager to avoid competition, of course argued that society would benefit by granting them an exclusive franchise to generate power (as well as a distribution monopoly) thus enabling the monopolist to build optimally sized generating units. This linkage of distribution and generation was always a mistake. There were, as we have seen, good arguments for the case that electric distribution initially had the characteristics of a natural monopoly and good reasons to encourage building this infrastructure by granting monopoly franchises. The argument that electric generation is a natural monopoly is less persuasive, but in fact was seldom if ever challenged. It was just assumed that generation and distribution monopolies were one—that the two went together. There are no reasons why a distribution monopoly can't purchase electricity from a variety of generators, and this separation was finally allowed in 1978 in the United States.

The 1978 Public Utility Regulatory Policies Act allowed non-utility generators to build cogeneration plants and not be subject to many provisions of the Federal Power Act or to state utility rate-of-return regulations. The Federal Power Act (FPA) is contained within the Public Utility Act of 1935. It gave the Federal Power Commission (later to be known as the Federal Energy Regulatory Commission) the authority to regulate interstate transmission and wholesale electricity sales. Under the FPA, electric sales from cogenerators and small power producers to a public utility would make the cogenerator, or small power producer, itself a public utility subject to PUCHA and state utility regulation.[30] The 1992 Energy Policy Act created a new category of electric producer, the "Exempt Wholesale Generator" (EWG), which is also free of these regulations. These law changes started a revolution that opened the market for new ideas in generation, which killed the optimal position of the large plants in a few short years.

However, full retail competition was not allowed anywhere in the United States before 1998 (even though there were early pilot projects covering one community in a state to learn what mechanisms might be used to open competition). From 1907 through 1978, granting exclusive generation franchises was assumed to provide economies of scale. Some states went so far as to ban on-site generation of power unless the power plant was owned and operated by the owner of the factory or office building. In all states, including those that allowed third-party ownership and operation of on-site power plants, backup power could be purchased only from the monopolist, and the monopolist was not required to purchase any excess power or to even operate in parallel with the on-site plant. These laws kept alternative power companies from developing or forcing the pace of innovation. Without examples of non-utility firms as successful generators, society gradually came to assume that electricity was so complicated and sophisticated that it had to be generated by the monopolist. Generators argued that with a monopoly they could build larger and more efficient plants, and, at least in theory, they would pass the savings of scale back to the electric consumers. The technology choices of 1930 seemed to support these arguments. Bayless shows that today's technology choices enable small generators to compete in the marketplace.

Generation Technology Primer

When Thomas Edison invented the electric light bulb in 1879 and commercialized electric generation in New York City, generation technology was in its infancy. The biggest machine yet invented to make power was a piston-driven steam engine, which produced about 200 kilowatts, or the power for two thousand light bulbs. These small machines required a separate boiler and converted only 8 percent of the total energy in the coal (which was the fuel) to electricity—the rest became heat. The first commercial power plant was on Pearl Street in the Wall Street area of Manhattan, New York. Edison recovered the steam left over from this early generator and piped it to nearby buildings to sell for heat in the winter. This generated more revenue and lowered Edison's net cost of making electricity. This combination of heat and power generation, or CHP, was the early pattern for all electricity generation and led to the development of district steam systems in most major cities. Until recently when there has been a spate of utility name changes, many of the utilities were named their major city name plus Edison and these all had steam systems to deliver and sell the waste heat. Examples include New York Edison (later Consolidated Edison), Detroit Edison, Southern California Edison, Boston Edison, and many others.

By 1900, steam-turbine technology began to eclipse steam-piston engines as the most efficient way to convert steam to electricity. A steam turbine is basically a windmill, with one or more rows of fan blades. High-

Fig. 6. Typical Steam Turbine[32]

Low Pressure Steam Out

High
Pressure
Steam In

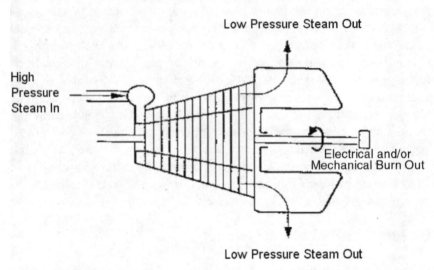

Electrical and/or
Mechanical Burn Out

Low Pressure Steam Out

pressure steam is piped to the first row of buckets that are simply sophisticated fan blades. The steam, traveling at a high velocity, hits these buckets and the impact causes buckets to move and thus force the shaft to turn. The shaft is connected to the rotor of a generator, which when turned, generates electric power. This process absorbs some of the steam's pressure and temperature.

Meanwhile, the steam has hit the first row of buckets and reversed its direction. A row of fixed blades redirects the steam in the direction the turbine is turning, whereupon the steam hits a second row of buckets, giving up more pressure and temperature, and providing more power to the shaft. Today's largest turbines may have fifteen or more rows of buckets, each larger than the last row to handle the expanding steam. At the back of the steam turbine, after the final row of buckets, there is still steam, but its pressure and temperature has been reduced. A portion of the energy has been removed. To increase the amount of energy removed, water from a river, lake, or cooling tower is used to cool the steam and create a partial vacuum; today's best turbines convert as much as 38 percent of the energy in the fuel to electricity. This process is called a *Rankine cycle,* named after William John Macquorn Rankine (1820–1872), a Scottish engineer and physicist.[31]

Waste heat becomes visible on a cold morning. Photo by Mark Parchman.

To understand how society can mitigate global climate change and not disrupt the economy, it is vital to understand one basic principle. The steam left over at the end of the electric generating cycle in even the best, most efficient steam turbines yet built still contains between 62 percent to 70 percent of the energy that was used to make the steam. This energy must be removed—dumped into some heat sink—in order for the steam turbine to work at all. A very modern power plant using only this Rankine-cycle technology can convert up to 38 percent of the energy in the fuel to electricity. The remaining 62 percent of the fuel energy remains as heat, which is usually dumped into a nearby river or lake, or is vented into the atmosphere.[33]

The picture above[34] shows a coal-fired power plant at Craig, Colorado, dumping 65 percent of the energy it produces into the atmosphere in a cloud of water vapor from the cooling towers. The standard way of removing this heat is to draw in water from a river or lake and cool the spent steam so it condenses to water. This transfers much of the energy to the water. This is why so many power plants were built beside rivers and lakes. An alternate approach, also widely practiced, is to reject the leftover

Fig. 7. Conventional Generation

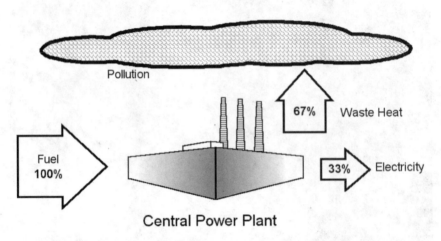

Central Power Plant

heat into a cooling tower, which evaporates water to remove the heat, and then sends the heated vapor into the air. It is this vapor condensing in the cold morning air in Colorado that you see in the picture above. Although monopoly-protected electric utilities have no incentive to recover and use this heat, a competitive power industry would stop throwing away this valuable heat and instead find ways to sell the heat to consumers.

Figure 7 depicts the typical approach to electric power generation all over the world. The numbers are U.S. average delivered efficiency in 1996.

The typical electric-only generating plant, built far from thermal users, uses 100 units of fuel to produce 33 units of delivered electricity. Two-thirds of the energy in the fuel ends up as waste heat, rejected into lakes, rivers, or cooling towers as in the picture above. The pollution is three times what it could be.

Figure 8 depicts an alternate approach—a better way to produce power. Instead of one large electric-only plant, a series of smaller "local" plants are built near thermal users—factories, universities, medical centers, and city centers. The same 100 units of fuel now produce three products: heat, cooling, and electricity. Up to 90 percent of the energy in the fuel can be converted to useful energy in these trigeneration plants and waste can be reduced to 10 percent. Generating electricity in one plant and heat in a second plant is inefficient. Why use two fires when we

Fig. 8. Combined Heat and Power (CHP)

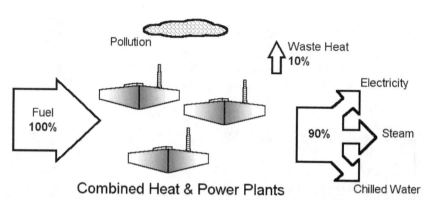

Combined Heat & Power Plants

need only one? Why use twice as much as is needed? Tomorrow, we must use one fire to produce heat, power, and when appropriate, chilling. This will save money and reduce CO_2 and other pollution.

Cogeneration plants now produce about 9 percent of U.S. power. They recapture some of the normally wasted heat and sell it as a second product. An added 10 percent of U.S. power comes from renewable energy, mostly hydro. Nuclear power produces 20 percent of the nation's electricity. Thanks to obsolete monopoly regulation, 60 percent of U.S. power comes from fossil fuel plants that throw two-thirds of the energy away as heat, producing needless pollution.

From the early part of the century on, steam-turbine technology was steadily improved, and Rankine-cycle plants used steam at higher and higher pressures. By 1960, the technology had reached maturity and stopped gaining efficiency. The best possible efficiency with steam-turbine technology alone seems to be about 38 percent, and employs highly sophisticated supercritical steam. Such technology is economic only in very large central power plants. The explanation plaque on the exhibit of a modern supercritical steam plant at the Smithsonian Museum of American History in Washington, D.C., explains as follows:

> In converting water to steam, a certain amount of the energy is used which can not be regained (as mechanical power). However, once the steam is formed, the energy which it derives from the fuel is propor-

tional to the steam's pressure. Thus, the higher the steam pressure used in an engine or turbine, the smaller the initial loss (relatively). The history of steam power, therefore, is one of increasingly higher pressures.

Today, the largest central-station steam generators are of the "once-through" type. They operate above 3,206 pounds per square inch, the critical pressure for water. At this point, the density of steam and water is the same when the temperature is above 705 degrees Fahrenheit, preventing natural circulation.

The degree of sophistication in these plants is very high, and no further gains in efficiency have been made over the past three decades. Transmitting this power to consumers through wires and transformers loses some of the electricity as heat and the net result is that, on average, only 35 percent of the fuel's energy reaches the user as electricity.

Figure 9 shows the average U.S. generating efficiency for electricity from the introduction of commercial electricity in 1880 to modern times. Edison's first generators converted only 8 percent of the energy in fuel to electricity. Efficiency then rose steadily to a peak of 33 percent, still wasting nearly two-thirds of the fuel. By the late 1950s, our society was fully electrified; there were many uses for electricity that had no competition from other energy forms, so people had to buy electricity. Monopoly protection insulated the electric utilities from competition with other electric generators, and regulators insisted that any savings from efficiency improvements be passed through to the electric customers. Faced with these rules and lack of competition, the utility industry stagnated and slowed the pace of improving its technology and efficiency. The graph shows that delivered efficiency of electricity from all U.S. thermally based generating plants has remained constant at 33 percent efficiency for four decades.

The obvious question is why the regulators have not demanded increased efficiency, and the answer is fundamental to how free markets work. **Regulators have asked their utilities to build efficient plants, but this urging has produced small progress.**

The driving force for change in almost every endeavor is a competitor who desires to gain some share of the business, who lies awake at night trying to imagine a better way, and who then risks a new approach or a new technology. Figure 9 shows that regulatory urging has been a very poor substitute for competition in producing efficient solutions.

Fig. 9. Energy Generation Efficiency Curve[35]

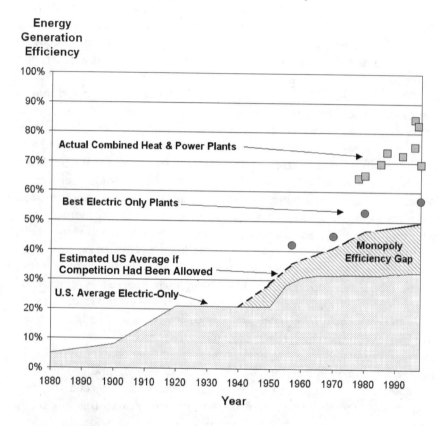

During these same four decades of stagnant average efficiency, technology has improved substantially, enabling electric power to be generated much more efficiently. The series of gray circles show typical efficiencies possible from the best technology that was available at each point in time, assuming the generation of electricity was kept separate and the heat continued to be rejected into the atmosphere. The area between actual efficiency and best technology is the theoretical efficiency gap, assuming that old plants are replaced as soon as new technology is developed. This is obviously not economic because of capital costs, and even in an intensely competitive world, it would take time for the older and less efficient plants to be retired. The most important tool for achieving greater efficiency is to combine the generation of heat and power, and to supply normally wasted heat from electric-only generation to homes, institutions, offices, and industry. The squares show the efficiencies at various times in the past that were obtained by the best power plants Trigen built. These plants combine heat and power generation. The difference in efficiency between these plants and the average U.S. generation efficiency is what we currently waste by relying on old, electric-only generation.

The thick dashed line is our estimation of the average generation efficiency that would have been optimally economic at each point in time. It assumes market forces were free to build new plants and deploy new technology whenever there was an opportunity to cut operating and fuel costs enough to earn a return on the required new investment. To achieve this level of average efficiency, we would have retired older generation plants when they were no longer economically competitive, and replaced them with a combination of the best available electric-only generation and with even more efficient combined heat-and-power plants. In many cases, competition would have led owners to modify the plants to recover heat, or to incorporate new cycles to increase efficiency. A substantial amount of dispersed generation would have been built near users of thermal energy where normally wasted heat could be recovered and sold. This estimate of what could have happened may be low, because, had there been markets for dispersed technology and for combined heat-and-power plants, there would have also been a more rapid development of the technology and a drop in the price of such plants due to volume production.

There is one more reason why our governments agreed to grant monopoly protection to generators, which then created this monopoly efficiency gap. In an era of no concern with carbon dioxide and before there was any natural-gas distribution, the ideal fuel for electric generation was coal. Coal is cheaper to handle at one large plant than at several smaller plants and this argues for large-scale power plants. Power-plant technologies before microcomputer development were very labor intensive. Operators were needed to manually adjust water-feed valves, shovel in coal or at least monitor the coal feed, remove ash, and constantly inspect various points of the boiler and associated equipment. (Today, all of these functions are automated and controlled by computers that allow one operator to monitor every function.) Power companies argued that they needed the certainty of monopoly electric sales to build the larger and more efficient power plants. The alternative was to have many smaller and less efficient coal-fired generators scattered around.

During the decades of the thirties, forties, fifties, and even early sixties, the utility industry placed all of its emphasis and research on creating bigger generating plants. Utility executives gathered every year to share ideas and tell stories. (The law did not allow them to be competitors.) The bragging rights at the Edison Electric Institute meetings of utility executives went to the executives from the company with the largest new plant.[36] The industry poured billions of dollars into research to make these large plants possible and to make them more efficient in generating electricity only, but by 1960 the Rankine-cycle technology was fully mature, and efficiency gains stopped. Until 1960 progress was steady and large engineering firms such as Stone and Webster, United Engineers, Ebasco, and others focused their technical talent on improving the Rankine-cycle power plant, making steam in a boiler and driving a steam turbine. The utility industry put little to no effort into gas turbines or combined heat-and-power plant development (CHP).

The Eighty-Billion-Dollar Monopoly Efficiency Gap

The area between the average U.S. thermal generating efficiency and the heavy dashed line is the *monopoly efficiency gap*. This loss of efficiency is the price consumers and the environment pay for continuing monopoly protection of this vital industry. The monopoly efficiency gap represents the gap between the speed of diffusion of innovations in free markets and the slower introduction in monopoly markets, as depicted earlier in figure 4, "Diffusion of Innovations." Because society has adopted the central generation paradigm and enacted a host of rules that unthinkingly assume all generation is central, the monopoly efficiency gap has been forced ever wider. The cost to the American public of the *monopoly efficiency gap* is $60 to $80 billion per year.[37]

More Cooperation, Fewer Innovations

The monopoly electric industry worked in cooperation rather than in competition, and this probably sounds like a good plan. Most people thought so, even as they witnessed stunning technological and economic progress in other industries that competed every day. For utilities, the deal was simple. Each had a franchise territory where all of the customers were captive. Antitrust regulation did not apply, so the industry worked together and shared research, rate-making approaches and marketing of electricity, and of course, lobbying for federal and state legislation to protect their monopolies. Older readers will remember "Reddy Kilowatt"—one of the most successful and widely known trademarks of all time, used to communicate the broad benefits of electric service.

There are undoubtedly gains to be made from such cooperation, especially for the cooperating companies. From time to time, our government encourages such cooperation, such as the computer-chip-making joint venture, Sematech. There are also very serious shortcomings to allowing an entire industry to cooperate and not allowing anyone to compete. Year after year, innovations that did not fit the single approach to power generation of the monolithic utility industry were missed, and more and more money was invested in outworn paradigms. By contrast, a competitive market holds promise of large rewards to the firm or individual who overturns the conventional wisdom and takes risks to deploy new and more efficient approaches.

I first saw this focus on output power at Cummins Engine Company in 1974, where four hundred engineers with advanced degrees worked at the technical center to try to increase the efficiency of diesel engines. Their goal at the time was to raise efficiency from roughly 33 percent to 33.5 percent, but it was seldom asked whether we could make use of the 67 percent of energy being thrown away in the exhaust and radiator as heat. To my great surprise, the giant electric generating plants that everyone assumed to be very efficient were also throwing away 67 percent of their energy as heat, and almost no one in the entire utility industry even asked whether they could make use of that heat. There has been little room in the neatly divided electric monopolies for breakaway behavior—no room for the "cowboys" and "mavericks" who change every competitive industry.

This ninety-year protective monopoly cocoon has resulted in the electric utility industry being less than half as efficient as it could be. The industry burns at least two barrels of oil, or at least two carloads of coal for every barrel or carload that it really needs. The industry overcharges its consumers by at least 30 percent compared to what appear to be competitive prices. Most important, this ninety-year protective cocoon has very seriously damaged the environment by allowing the industry to burn too much fuel and thus to emit excessive pollutants and carbon dioxide. The resulting needless acid rain is leading to widespread death of entire populations of dogwoods, red spruce, maple, and many other trees.[38]

This protective cocoon surrounding electric generation and sales all over the world has greatly exacerbated the needless emission of carbon dioxide and is responsible for societies' excessively rapid burning of their fossil fuel reserves. Monopoly regulation of electricity explains much of the current excessive emission of carbon dioxide. These large claims will be discussed in more detail later, but first we must try to understand and explain why there has been such an emphasis on fossil fuel and why monopolists fail to innovate.

ALL MONOPOLIES ARE SLOW TO INNOVATE

Fresh thinking in business normally comes from one of two motivations—greed or fear. People employed in competitive enterprises are motivated by their own self-interest to seek growth of market share and profit at the expense of other firms. As a result, competitors must constantly look for new approaches and fresh ideas, or run the risk of losing market share to others and going broke. In every competitive business, those who lag in finding and implementing better and cheaper ways to serve customers find their market share shrinking and their profits falling, at which time a more powerful motivation takes over—fear.

Fear is the great cleanser of the status quo. The world is always moving toward entropy—becoming less organized and less efficient—and it takes much continuing effort and rethinking of basic beliefs just to stay in place. Fear of failure causes competitive business organizations to rethink everything that is sacred. Fear is the "necessity" that mothers many inventions.

You Get What You Reward

The managers of protected electric utilities all over the world have been sheltered from both of these motivations for decades. The firms enjoy a complete monopoly of electric generation and sales, and if they take risks to innovate and reduce the costs, the public service commissions have historically made the company give 100 percent of the savings to the consumers. Greed does not pay. Why take risks? Growing larger by merging with or purchasing other electric utilities has also been severely limited by the restrictions of the Public Utility Holding Company Act

(PUHCA) of 1935, which makes it extremely difficult for well-run utilities to take over less well-run firms in the utility business. Without PUHCA's restrictions, an electric utility that was well managed could identify another electric utility that was poorly managed and offer an above-market price for that utility's shares. Then, when the well-managed firm gained control, it would institute better management practices, improve the results of the target utility, and be rewarded with enhanced profits. However, this has not been legal until recently. PUHCA limited each electric utility's potential growth in the electric business to activity within its franchise area. The utilities have focused on selling more electricity, and on serving other energy uses like heating with electricity, and even on helping promote economic growth in the towns and states the utility serves. However, real growth by expanding geographic reach, or serving customers outside the franchise territory, has been historically prohibited. Until the passage of the Public Utility Regulatory Policies Act (PURPA) in 1978, no one but the monopoly utility was allowed to generate and sell electricity inside its territory.

"NATURAL MONOPOLY" CHARACTERISTICS OF ELECTRIC GENERATION ARE GONE

There are no longer any "natural" monopoly characteristics associated with electric generation. There are no longer good reasons to prevent anyone who wishes from entering the electric-generation business. While such restrictions made sense during electric power's infancy, the technology of generation is widespread and easy to master. In fact, we have millions of self-contained electric-generating plants. The worldwide production of vehicles was 10.6 million in 1950 and has risen to 54 million in 1997.[39] Every one of these automobiles, trucks, and buses has a self-contained electric-generating plant to provide electricity for controlling the engine and operating the lights, radios, and fans. Assuming an average life of eleven years, the world has about five hundred million operating vehicles, each with its own electric generating plant. No one gives a second thought to operating his or her own vehicle without an extension cord connected to a regulated monopoly supplier of electricity.

Self-generation, by contrast, is seen as unreliable or uneconomic, but the evidence does not support this view. Every hospital, by law, has an on-site generating plant sized to provide emergency power in case the electric grid has an outage. Every ship and airplane has its own generating plant. With nearly twenty-eight thousand merchant ships operating in 1995, each with its completely independent electric-generation plant, the world does not lack for experience or equipment supply of dispersed generating plants. However, by and large, very few people or businesses have given much thought to operating their own electric plant, or having some third party install and operate such a plant.

The technology of small-scale generation exists and is quite mature. Technically, all of the stationary generating plants can connect to an end user of electricity or to the grid of wires, and sell their product—electricity—in a competitive market. What is more important is that there is always heat left over when burning fossil fuel to generate electricity. This heat (normally wasted) can be sold or used as a second valuable product. Selling the normally wasted heat means getting another useful product from the same fuel, thereby reducing the cost of the electric generation. It is not difficult to capture and sell this heat, provided that the electric generation plant is close to the heat user. However, the slavish adherence to the monopoly concept of building ever larger generating plants, often located far from heat-using population centers, has denied consumers the competitive possibilities of dispersed power generation for the past seventy-five to ninety years.

Eliminating the prohibitions against generation and sale of power should also eliminate the present monopoly of distribution. There are many places where local combined heat-and-power generation would be more efficient if the owner were free to run electric wires to nearby electric users and bypass the regulated grid. If this lowers the cost of electricity for other consumers, versus the existing grid, why not allow the competition?

PUHCA PROTECTS BAD MANAGERS

In the early part of the century, Samuel Insull built a huge network of electric holding companies that were each highly leveraged with debt. To stretch his own capital, Insull would use one company to control the next, thus increasing financial leverage. Leverage works in both directions. When everything is going well, financial leverage multiplies the profits to equity, but when things go badly, leveraged structures are vulnerable to bankruptcy. Insull's thinly capitalized structure was vulnerable to failure in an economic downturn. The Great Depression wiped out Insull's empire in 1932. Congress reacted by enacting the Public Utility Holding Company Act (PUHCA) in 1935, which severely restricted the ability of one utility to own another utility. At the time, there was a fear of repeating the Insull experience, although I seriously doubt it was possible. By then, the technology of electric power was freely available, capital markets were ready to back power companies, and the "magic time" had passed. I term "magic time" as that dreamtime for entrepreneurs when they can do no wrong—when their strategy is clearly superior to all competitors. These moments are few and short-lived, because of the most damnable force ever unleashed against entrepreneurs—competition.

PUHCA was probably unnecessary when it was passed; it was merely the "lymph node" that swells up after the body has defeated the infection. What is clear is the damage PUHCA has done and continues to do. It makes it very difficult for anyone to take over an electric utility, no matter how badly that utility is managed. By contrast, competitive businesses are constantly subjected to raiders. These raiders are market cleansers who search for companies whose assets are underutilized or poorly managed and whose stock price is low. When raiders strike, they offer the existing shareholders a 20 to 40 percent instant premium over the current stock price and then purchase all of the stock. Inefficient management is expelled, low-value strategies are discontinued, and the new owners attempt to create value with a different vision. The fear of being raided keeps the management teams and boards of most companies alert and focused on serving customers and creating value for shareholders.

Utility managers, until recently, were exempted from this fear because they were protected by PUHCA. The law was passed to prevent a repeat of the depression experience of utility companies going bankrupt. Ever since, the consumers that the law was supposed to protect have been forced to pay higher-than-competitive prices, sometimes receive poor and surly service, and suffer from dirty air. Examples of poorly managed utility companies abound. I challenge readers to find anywhere in the competitive sector of our economy examples of such egregious bad management as Long Island Lighting Company (LILCO), which spent nineteen years and $6 billion on Shoreham Nuclear Generation Station, and never produced a kilowatt. Without PUHCA protection, some better-managed firm would have raided LILCO and installed better management long before the passage of nineteen years or the wasted expenditure of $6 billion, and Shoreham is only one of LILCO's many management failures.

PURPA INDUCES MANAGEMENT CHANGES

One of the most important impacts of the 1978 Public Utility Policy Regulatory Act (PURPA) was unintended, as far as I can tell. PURPA allowed non-utilities to generate power, if their generation met efficiency standards of roughly 45 percent fuel converted to useful energy. This standard is effectively one and one-half times the efficiency of the national average electric utility efficiency (33 percent delivered efficiency). The unintended consequence that has turned out to be significant was that PURPA opened the door for the electric utilities to grow outside their monopoly territories. With the possibility of growth, the self-interest or greed motivation came back into play in a stultified industry. Over fifty of the monopoly electric utilities formed active, unregulated, independent power producers (IPPs) to build and operate generating plants in other utilities' territories. Some of these utilities' IPPs have expanded all over the world and are taking advantage of the move to liberalize electric generation wherever it occurs. This has changed the dynamic of the utility business profoundly, albeit slowly. Prior to PURPA, electric utility growth was severely limited. Highly entrepreneurial-type people like those who developed the electric industry do not typically find a mature, growth-constrained monopoly an attractive place to work. Prior to PURPA, there was no utility equivalent to "Silicon Valley" for computers. The main job of the CEO of every monopoly utility was to preserve the monopoly, win regulatory approval of rates that would keep cash flows high, and pay out a very large portion of the earnings every quarter as dividends.

This is not to say that many utility management teams did not try to grow their businesses in other areas. The regulated investor-owned electric utility companies in the United States have collectively invested billions of dollars in nonelectric ventures, in general with very poor results. Utilities used their strong cash flow to purchase real-estate developers, used-car dealerships, hotels, and many other unrelated businesses during the years 1970–1985, but largely stopped the practice due to poor results. Managing a monopoly that has little chance for significant growth and little chance to fail requires some important skills, and I have much respect for the individuals who rose through the ranks and became CEOs of multibillion-dollar electric and gas utilities. However, the skills needed to succeed in a competitive business are very different from the skills needed to succeed as a monopolist.

These top managers of protected monopolies, blessed with rate-making policies that pass through all costs, have often fallen into the trap of convincing themselves that they are competitive business leaders. Because they must convince the regulators of their outstanding management, their most critical skill is to put a spin on all actions taken that will satisfy the regulators. Most utility CEOs have become quite good at this essential skill. Persuading customers to choose your company when there is freedom of choice requires different skills, and the utility management teams have sometimes failed to appreciate this difference. After years spent persuading utility commissions that all expenses are necessary, it is no longer possible to tell what is an essential expenditure and what is pure fat, or worse, a drag on efficiency. Humans are all capable of endless self-delusion, until it is checked by reality. The self-delusion reality-check in business is competition, and the monopolists are denied this painful but vital soul cleansing. They inevitably grow soft on cost control, and err on the side of safety and reliability.

All of this helps to illustrate how the electric utility industry became stultified over the many years of protective regulation prior to PURPA in 1978. The original entrepreneurs who built electricity from a concept to one of the truly important businesses in every country are long since gone—dead and buried. The aggressive organization built by Samuel Insull took no prisoners and grew every way possible. His excesses led to PUHCA in 1935, but he sped the development and availability of electricity. Forty years of regulated and protected monopoly prior to PURPA deadened that early entrepreneurial culture throughout the industry and replaced it with a risk-adverse, caretaker culture.

PURPA changed this landscape. It became possible for a utility to re-invest some of its cash flow in a business it understood—generating electricity. It became possible to achieve top-line revenue growth by generating and selling electricity outside of the franchise territory. This improved opportunity for rapid growth probably impacted many career decisions. Aggressive young managers decided to stay in the electric business. Utilities have always been a good place for a young technical graduate to go to work. With no real pressure on costs (which are all passed through to customers) the typical utility has strong training programs, abundant specialists, and plenty of time to slowly and carefully work through decisions. In many regards, this is a good environment for a young and aspiring technical manager. But after polishing engineering classroom skills and gaining some practical experience, a career in the traditional regulated utility typically has less challenge and growth opportunity than is available in many competitive businesses. As job security became less important in the 1960s, the most highly motivated people (who can be the hardest drivers and innovators) often left the utility early in their careers to search for more challenge. Those who remained and ran the electric utilities in the sixties and seventies were more risk averse. It reinforced a culture of careful evolutionary development. As these risk-averse personality types gained power, the monopoly gap grew wider. The difference between the optimal approaches and the average efficiency grew as a result of year after year of uninspired management of electric utilities.

PURPA opened the door for some young Turks to stay in the company, but to work for the utility's Independent Power Producer (IPP) subsidiary, where the currency was new ideas, new market deals, and most important, freedom to profit from efficiency gains. In a short five years Jim Pignatelli built an IPP business for Southern California Edison that produced $90 million in after-tax profits. Similar large IPP subsidiaries have been built by CMS of Michigan, Entergy (based in New Orleans, Louisiana), Duke (based in North Carolina), and Louisville Gas and Electric (serving Kentucky), to name a few.

Some utility-management teams resisted, came late to the game, and did not allow PURPA to change their conservative cultures; their non-competitive culture today is their own fault. PURPA gave each electric utility twenty years to move toward more competitive behavior, to attract more aggressive managers, and to hold onto entrepreneurial new hires who were oriented toward high growth and competitive challenge. Today, aggressive companies like Cinergy are being driven forward by people like Jim Rogers, who was recruited from the deregulated interstate gas business to take over a nearly bankrupt Public Service of Indiana. Rogers worked through the hard decisions to write off a failed nuclear plant (Marble Hill), cut costs, and then helped engineer a merger with Cincinnati Gas and Electric to form Cinergy. Cinergy went on to

purchase half of an English electric distribution company (Midland's Electricity Board), and to enter the combined heat-and-power business with
Trigen Energy Corporation. Rogers will grow his business, or die trying.
He has energized a staid energy company and is breaking down the stul-
tifying impact of monopoly protection.

Charles Bayless put himself through school to earn bachelor's and
master's degrees in electrical engineering by serving as a lineman for
West Virginia Power. He went on to earn both an MBA and a law degree.
Bayless is a bright, motivated individual who chose the electric utility
business as a career. In 1981, he was hired as chief financial officer of the
troubled New Hampshire Public Service (PSNH). Cost overruns on the
Seabrook nuclear plant were forcing bankruptcy; the company had not
anticipated government actions that made Seabrook's completion economically impossible. Bayless learned at a deep, gut level that electric
utilities could fail and that the regulatory cocoon allowed problems to
become too large to be solved. After steering the PSNH bankruptcy in a
manner universally admired by the many at-risk financial institutions,
Charles Bayless went on to become the chairman of Tucson Electric
Power, another turnaround situation. While restoring financial health to
TEP (now UniSource), he has been one of the industry's earliest and
most outspoken advocates of deregulation—of opening the electric
market to all and removing all government subsidies to federal and
municipal power companies. In July 1998, Bayless was tapped to head Illinova Power headquartered in Decatur, Illinois.

I agree with utility analysts and others in the power industry that
there are now great gaps in management quality and strategy among the
200-odd investor-owned utilities. Even though PUHCA has not yet been
repealed, its prohibitions against merging with other electric utilities are
being less vigorously enforced. Many mergers have occurred or been
announced. Within two to four years of the anxiously awaited repeal of
PUHCA, there will be a great consolidation of electric utility companies,
and the weak managers will be retired. This alone will increase the
industry efficiency and reduce fuel use, costs, and pollution. It will
shorten the time for diffusion of innovations to be implemented in the
industry, and most of those innovations involve burning less fossil fuel
and thus emitting fewer greenhouse gases.

PROGRESS IN TECHNOLOGY

Diesel-Engine Progress

World War II spawned two important technologies that could be harnessed for power generation. The diesel-engine business, struggling since its U.S. introduction in the 1920s, received a shot in the arm in the form of military demand for rugged engines that could operate on diesel oil. By 1960, engine technology was improving rapidly. Diesel and gas engines were almost as efficient at converting fuel to mechanical energy as the huge central electric power plants. These high-speed diesel engines, made notably by Cummins Engine Company, Caterpillar, and Detroit Diesel, became more efficient in converting fuel to mechanical energy than the average large power plant.[40]

These engines could run on gas or oil and be located at a user's facility where the normally wasted heat could be used to eliminate the need to burn other fuel for heat, and this would reduce costs and pollution. Not surprisingly, the deployment of these mass-produced power plants represented a great threat to the monopoly utilities, and they fought such deployment very effectively. Efficiency is not the monopolist's main goal; preserving monopoly rents is. What logic would persuade government officials of the need to preserve monopolies if renegade developers of small-engine power plants were allowed to prove what is true—that making power is as easy as owning a truck engine? Would-be power entrepreneurs have faced incredible efforts by the protected monopolists to prevent anyone from demonstrating that making power could be as easy as buying a truck engine. The public and the environment have been the losers.

As one small example of these efforts to preserve both the monopoly and the fiction that a monopoly is needed, look at the response of Consolidated Edison to the miniscule efforts of the five-person Cummins Cogeneration Company in 1977 through 1980. Mike Weiser and I formed the division for Cummins Engine Company and sought to place diesel-engine-driven generator sets in New York City office buildings and manufacturing plants. Con Ed assigned a senior vice president, Bertram Schwartz, to stop us. Among the many initiatives Con Ed took was a request to New York City to hold an extended public hearing into the health effects of cogeneration. Imagine a study to determine if recovering heat from mass-produced truck and bus engines would make people sick. Heat from a bus engine is used to heat the bus, just the same as we proposed to do, except that we would capture all of the heat and warm entire buildings.

A deputy mayor headed a task force to look into the Con Ed request and was told by the senior Con Ed environmental scientist, Dr. Peter Freudenthal, that if cogeneration plants were allowed in the city, the citizens would suffer from "pharyngeal dryness." He presented graphics to show that the exhaust from these engines would come out of the exhaust stacks at the top of tall buildings, drop directly to the ground, and be inhaled by the folks on the street. It was never clear why exhaust from our diesels would travel directly to the ground, but exhaust from the boilers we replaced would not. Neither was it clear why the exhaust from Con Ed's many New York City–area generating plants was not following these newly revealed laws of physics. Nonetheless, Mike and I had the unique experience of appearing at city hall before a blue-ribbon commission chaired by a deputy mayor to debate Con Ed's senior staff, who had retained a major public-relations firm paid for, of course, by the electric ratepayers. The city decided the request was without merit, but Con Ed continued to publish learned papers on the environmental dangers of cogeneration. The true danger of our small cogen plants was to the logic of monopoly protection. The time delays caused by such actions by the monopolist were a killer to fledgling companies trying to innovate in the power business. These small companies seldom had the cash flow to ride out the long delays.

I experienced another lesson about how monopolists maintain their position when the New York State Senate held a hearing on cogeneration and the senator in charge asked Bert Schwartz, executive vice president of Con Ed, and me to debate. Mr. Schwartz testified first, saying that, "If an electric user puts in his own cogeneration plant and leaves the Con Edison system, we will be forced to raise the rates to all of our remaining electric customers."[41]

When it was my turn to testify, I exhibited my youth and naivete by opening with a tongue-in-cheek response to the senator leading the hearing. I asked to be excused immediately so I could rush back to Columbus, Indiana, and meet with senior Cummins Engine Company executives to explain this new marketing strategy. I said that Cummins had never tried to sell engines by telling the big trucking companies that,

"If you do not buy your diesel engines from us, we will be forced to raise our engine prices to other trucking companies, most of whom are your competitors."

The senator was not amused. He stopped the hearings and gave me a ten-minute lecture on how Con Ed was forced by law to run wires to all of the remote rural customers in their territory (rural Queens, I assumed). He went on to tell me that I was out of order in his hearing to make fun of this fine company. After the hearings were over, two Senate staffers sought me out to apologize for not "marking my card." They said they should have told me that Con Ed was a major contributor to the senator's election campaign fund. Nearly every electric utility management mounted similar strong efforts to prevent entrepreneurs from installing cogeneration, even though such plants easily achieved double the efficiency of conventional generation.

These tactics make amusing reading nearly twenty years later, but they were brutally effective. In spite of Con Edison's very high rates, little cogeneration was built in New York City in the past two decades. Independent power companies succeeded elsewhere, but found little success when pitted against a truly determined monopoly. The combination of laws, attitudes, "monopoly-friendly" officials, and sheer economic staying power doom an alternative electricity provider. The playing field is not level.

GAS-TURBINE PROGRESS

Monopoly protection of the electric industry has also been rendered obsolete by the advances in gas-turbine technology, which were stimulated by military needs for airplanes and by civilian demand for air travel. The best gas turbines in 1970 achieved just over 20 percent efficiency and were very polluting. Today's gas turbines achieve from 42 to 45 percent efficiency and produce such small quantities of regulated pollutants that the plant operators have had to upgrade the sensitivity of their pollution-monitoring equipment to laboratory standards. A 1.5-megawatt gas turbine now being endurance-tested by a company named Catalytica, using their new Xonon combustor, is so clean that it is difficult even to measure the nitrous oxides or carbon monoxide emissions. Gas turbines in all size ranges can be mass produced, lend themselves to easy heat recovery, and can be the prime mover or engine to drive generators in dispersed power plants. These plants can be owned and operated by many competitors. Gas-turbine-based combined heat-and-power plants are in operation today that achieve up to 91 percent efficiency, can earn a profit selling electric power at discounts of 30 to 40 percent, and reduce pollution by over 90 percent. It should not surprise anyone that monopolists have not been rushing to deploy this technology. This technology makes obsolete the economy-of-scale arguments for continuing monopoly protection of electric generation.

Fig. 10. Generation Efficiency of Various Technologies Today

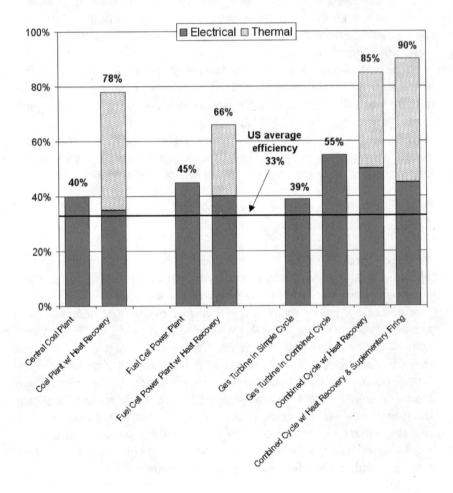

Figure 10 shows typical efficiencies for various electric-generation technologies commercially available in 1998. Note that every technology available today has higher than the United States average of 33 percent delivered efficiency. New combined cycle gas turbines are 55 percent efficient, but with heat recovery they can reach 90 percent efficiency. This is nearly three times the average. Open electricity to competition and these very efficient plants will be built, replacing the many obsolete plants that remain in service today. *The way to mitigate climate change and save money is staring at us from this chart.*

MONOPOLY PROTECTION LEADS TO EXCESS ELECTRIC USE

We began by noting that cooperative behavior among monopoly utilities appeared to have some advantages, but also had the seeds of many failings. The ninety years of monopoly protection have blocked the incentives for true efficiency in the energy-conversion industry. The enforced lack of competition has resulted in our great energy companies not doing what every competitive enterprise has been compelled to do to survive. Competitive enterprises must learn how to make what customers want. Monopolies are comfortable selling what they know how to make.

Although they often don't realize this truth, customers of the electric utility do not particularly want electricity. They want light. They want to be warm in the winter and cool in the summer. They want fresh air in their buildings. They want elevators that work—pumps and fans that turn. They would love to hire an energy company that excelled at supplying their true needs. In every competitive industry, suppliers neglect customer needs and desires at their peril. Some other firm comes along and says, "We will make what you want. We will remove the hassle you have of trying to produce some product that is not your core business. We will do it all cheaper than you can. And we will earn a profit."

As computer-chip technology made it feasible to build small personal computers, IBM ignored the change and sought to continue selling what they knew how to make—mainframe computers. Apple learned how to make something many customers wanted to buy—a personal computer—and grabbed a nice piece of the computer business. Later, IBM created a PC of its own, but others made IBM clones that were faster and cheaper. By the time IBM adjusted, its dominant position as computer maker to the world was gone.

Competition then attacked Apple. Apple tried to trap all the profits and keep any other company from building its wonderful designs that were significantly better than the competing "IBM clone" PCs. Once again, the customers' true desires were ignored. Commercial customers did not want to be dependent on a single hardware supplier, and opted for IBM–type machines that were manufactured by numerous companies. Academics and students have loved and embraced Apple, but businesses never really allowed Apple's single-supplier approach to dominate their market. Competition is nasty, unforgiving, and swift. No one wants competition against his or her business, but it is the engine that drives the development of better products and services and keeps consumer costs down.

To illustrate the point of how monopolists have not learned to make what their customers want, consider building services. Building managers hire security firms, maintenance firms to manage their office cleaning, and food-service firms to run their cafeterias. They hire others to provide all of these noncore activities and concentrate their capital and resources on the core real-estate activity. Yet, when it comes to air conditioning and heating, most buildings buy, install, and operate these devices. Why do building managers put capital and resources into the relatively complex task of converting electricity into chilled water to cool an office building, but outsource office cleaning and food service? Building managers often operate heating and cooling equipment in suboptimal ways, using needless extra energy, causing unnecessary pollution, and costing too much. In a competitive market, suppliers would offer these services. Where are the competitors? The electric and gas monopolists can't be bothered with such details, because they have the building's electric sales guaranteed anyway. Other potential service companies have found it difficult to gain a foothold in the energy-conversion business, thanks to monopoly protection. Competition would create new opportunities for energy producers to save money and reduce pollution by offering to install combined heat-and-power plants and providing the heat, electricity, and chilled water needed from one plant and one fuel. If many firms competed to supply the building with electricity, surely one or more would try to win the business by going beyond the electric meter and providing the comfort and light the customer really wants.

From 1930 through 1997, none of the fifty states allowed competition in electricity. As a result, society has no immediate experience with electric competition, and therefore, no knowledge of what is missing. This is changing. Four states (California, Massachusetts, Rhode Island, and New Hampshire) set 1998 as the starting point for opening electricity to competition. Six other states passed laws to open their markets in the years ahead. In 1998 and 1999 citizens in those states will start to see what they have been missing. The transition to competition in these states involves a payment to the former monopoly utilities for their "stranded" costs—their uncompetitive generating plants and electric-purchase contracts. These plants were state of the art when built, but by today's standards are inefficient and more expensive to operate than new power plants. The no-longer-competitive plants were built with the full blessing of regulatory commissions, and the utilities relied on monopoly laws to make the decisions to invest, so it is argued that the public must pay for the remaining book value.

When an industrial firm invests in a production plant, and some years later technology improves and renders the original plant too expensive to operate, that original plant is closed, sold, or significantly modified, at shareholders' expense. The customers have no responsibility to pay for their supplier's mistakes. Because the monopolist had an obligation to serve and made decisions to build with the full blessing of the state regulatory commission, it is generally accepted that the utility shareholders are not responsible for the reduced asset values due to competition. The argument is made that the utility relied on monopoly laws that are now being changed, and that it is society's responsibility to protect firms against changes in law. Consequently, there will be a surcharge or transition fee on all electricity sold for four to eight years after deregulation in every state. This will delay real competition, so little change is expected for the transition period. Then competition will begin in earnest. My prediction is that if governments end all barriers to efficiency, we will see a 40 percent drop in real electric prices within ten years, fuel prices will collapse from reduced demand, and all pollutants, including carbon dioxide, will be significantly reduced. The falling electric prices will lower the costs of everything else that is produced with electricity, and we may enter a long period of deflation, all because governments unleashed market forces. It is time.

3

Noblesse Oblige—
For Planet Earth

ON BECOMING GOD-LIKE

My mother speaks no French, nor would she consider herself a philosopher, but she repeated "noblesse oblige" to me many times as I was growing up and again later when we have talked about responsibility. To Mom, noblesse oblige means simply, "To whom much is given, much shall be required." It has always been her way of conveying a sense of ethics. Since much has been given—a sound mind, a good education, valuable experience—then much is also required. The powerful ethical message wrapped up in this simple phrase is one that humankind must collectively embrace before we destroy Earth.

What "has been given" to humankind is incredible power over nature, thanks to science and the scientific method of wresting truth from nature. We have discovered many of the laws of nature, discovered much about the universe around us, and gained an awesome degree of control over nature. We have in fact acquired many of the powers formerly thought to be the sole domain of the gods.

In Homer's *Iliad,* the classical Greek deities had supernatural powers, particularly over human life, but were severely limited by the relentless force of fate. The gods were most important in their role as guardians of the city-states and could provide information, through divination rites, about one's future on earth.[42] Today, scientists eschew divination rites, but apply knowledge to predict the positions of the planets, eclipses, the weather, and a great deal more. Prediction by visiting the oracle was a myth that everyone wanted to believe, whereas the predictions of science can have near-perfect accuracy.

Compare the power of the Greek gods with humankind's power over nature today. Achilles,[43] the hero of Homer's *Iliad,* was born a mortal but his mother, Thetis, dipped him in a magic river. This gave Achilles immortality wherever the water touched his body. Thetis held Achilles by the heel, so that heel did not become immortal, and injury to that remaining vulnerable spot finally killed Achilles. In the past two hundred years, medical science has discovered and gained mastery over many of the magic powers of this river of Greek myth. Doctors now protect us against diseases that ravaged our ancestors. Polio, diphtheria, smallpox, and bubonic plague are almost scourges of the past. Contagious diseases like typhoid were largely wiped out by modern sewage systems. Doctors regularly replace damaged or failed organs, repair the bodies of wounded soldiers or accident victims, and in many other ways give us partial immortality. We have gained god-like powers.

We also seem to have an Achilles' heel, which may be associated with another Greek concept—*hubris,* defined as "overbearing pride or presumption; arrogance." McGeorge Bundy, educator and special assistant for national security to presidents John F. Kennedy and Lyndon B. Johnson, may have captured the idea of our Achilles' heel when he said, "There is no safety in unlimited technological hubris."[44]

Human history is filled with examples of rapacious behavior and thoughtless use of newfound powers to wipe out entire ecosystems and species. Consider the slaughter of the American buffalo herds by people amusing themselves on train rides. Consider the dodo—a friendly little tasty bird that now exists only in sailors' diaries.

As medical science has improved, population has ballooned. Not yet smart enough to avoid fouling our own nests, we overfish the oceans, overwork the soil, and overcut the forests. One terrible example of this uncontrolled exploitation of natural resources is Thailand. Deforestation delineates most of the border of Thailand on an infrared satellite map. Humankind's Achilles' heel may be our capacity for a hubris that assumes we can solve any future problem even though we lack full understanding of the planetary consequences of our actions. There is a real danger that we will destroy the planet we live on before we understand how much we do not know. Global warming and acid rain are both threats to the planet but we underestimate the problems, denigrate the science, fear for any economic disruption, and finally pass the problem off to the next generation. Noblesse oblige?

The scientific method has been largely responsible for expanding knowledge and increasing human power over nature. Scientists test and retest every hypothesis, and subject all of their claims to peer review. Science refuses to accept research results until there is corroboration by other scientists. Science extracts nature's secrets with a constantly challenging group mind. It has not always been this way. Knowledge used to be controlled by high priests or government leaders who maintained authority over matters of philosophy and views of nature, while artisans carefully protected their knowledge and passed it from father to son. In medieval times, theory was considered to be the highest intellectual endeavor, and it was not considered necessary for intellectuals of the time to confirm their theories with experiments or to have their theories subject to review by others.

Humans were sacrificed to placate the gods or to insure that the pharaohs would have a good afterlife. As late as 1900, we still thought God sent the fevers that killed men working on the Panama Canal to punish these men for leading sinful lives. Only in 1905 did Dr. William C. Gorgas of the United States Army Medical Corps seek to wipe out yellow fever and malaria by destroying the kinds of mosquitoes that carry these diseases. Yellow fever was eradicated by 1906, and the rate of death caused by malaria was significantly reduced by 1913.[45] In fact, the pots of water around the porch pillars protected those supports from termites but also served as breeding grounds for the anopheles mosquitoes, which carried yellow fever. If it was an angry God who killed the workers, she was apparently placated by the installation of screens.

New knowledge came very slowly during most of human history because so many hopelessly wrong theories were accepted dogma and were not subject to testing and proof. Galileo Galilei (1564–1642) was the first to use a telescope to study the stars (1610), and his observations made him an outspoken advocate of Copernicus's theory that the Sun forms the center of the universe instead of the Earth. Galileo's reward for advancing knowledge was persecution and imprisonment by the Inquisition in 1633.[46]

Nepal may be the most wonderful place one could ever visit, even though it is one of the poorest countries in the world. With a terrain that frustrates communications, many people live fairly isolated from modern ways. The mountain people have no tradition of the scientific method, and suffer as a result. On a recent trip, my wife and I visited our friends Kedar and Munu Thapa in Kathmandu, observing firsthand some health practices Munu works to change in her role as a women's health counselor. She showed us slides of Nepalese peasant neonatal practices that helped to explain an unusually high infant mortality rate. Remote mountain families believe the baby should not be clothed until it is named—roughly seven days after birth. In frigid, closed rooms, peasants burn dried cow dung in an open brazier next to the naked infant. The resulting build up of carbon monoxide, generally tolerated by the fully developed lungs of the parents, asphyxiates many babies. Lack of scientific knowledge kills people, and always has.

The developed world has come a long way in the 365 years since Galileo's imprisonment for challenging the church's dogma with observations and facts. Human knowledge has doubled and redoubled as this amazingly effective scientific method has extracted secret after secret from nature. We have learned to fly, and have even flown to the moon. We have discovered and managed to control nuclear fission, part of the engine that drives the stars and that provides the energy for all life. We dream of controlling the main mechanism for stellar energy, nuclear fusion. We have also learned to extract fossil fuel from the earth, convert it to useful energy, and in the process greatly increase our standard of living. We are, in fact, on a course to burn all of the fossil fuel produced by plants and sunlight over four hundred million years in just a few hundred years. We are gluttons for fuel, using much more than we need to meet our energy requirements.

Knowledge can be used to produce good or bad results, sometimes-unintended bad results. In *The Heat is On,* Ross Gelbspan captures the deepest danger inherent in humankind gaining God-like power.

> With the growth of the human population to five plus billion and with the development of extraordinarily powerful technologies, we ourselves have become a force of geological magnitude... the massive emissions from our factories, cars, and homes are altering the very seasons of the earth and the balance of its most fundamental cycles of life.[47]

As we have learned to harness energy, our standard of living has increased. Fred Cottrell, professor of government and sociology, Miami University, Oxford, Ohio, charted the relation between energy, social change, and economic development in 1955 in a landmark study titled *Energy and Society*. His thesis was "that the energy available to man limits what he can do and influences what he will do."[48] We start with only our own muscles, and are lucky to even survive until we harness fire to keep us warm. After millennia of having only fires and use of work animals to supplement its muscles, humankind figured out how to make windmills and waterwheels and the standard of living increased. When James Watt improved on the earlier Newcomen steam engine invented by Thomas Newcomen in 1705, and made it practical to use steam engines,[49] he enabled man to capture mechanical energy from fire and the standard of living started to grow in earnest. Each further increase in the standard of living involves harnessing more energy to our service.

The villager in Nepal without electric light bulbs understands Cottrell's thesis at a very fundamental level. Clearly, harnessing energy has produced a plethora of benefits for humankind. Much has been given. But if we are to survive, much is also required.

Cottrell distinguishes between energy use, which is the use of heat or light, and energy conversion, which is the conversion of the energy content in fuel to heat or light. He writes, "One of the problems . . . will be to discover some of the conditions under which man will be likely to continue to pursue a course demonstrably 'wasteful' of energy in preference to a more efficient way. Man seems frequently to follow such a course."[50]

Society today is in one of those times—"demonstrably wasteful of energy."

TOO RAPID ENERGY RELEASE

We are rapidly liberating the carbon that was laid down by solar energy over a half billion years. We have already changed the chemical composition of the global atmosphere, increasing the carbon dioxide concentration to levels not seen in 160,000 years, and the level will double in another seventy years if we do not reduce fossil fuel consumption. Science tells us that this increased carbon dioxide concentration will trap infrared radiation from the surface of the planet, raising temperatures.

Science also shows that this rising temperature could increase evaporation and produce more cloud cover. Clouds are much more effective than carbon dioxide at trapping infrared radiation from earth. We have all experienced this effect. On a winter night, if the stars are out and the sky is clear, local temperatures will drop much more overnight than if there is cloud cover. We have the science to understand that the global weather is driven in very complex ways by the heat of the sun, and that small alterations may produce feedback effects that magnify local weather changes. Scientists studying global climate predict that increased global average temperatures will: (1) produce weather extremes of hot and cold, and (2) increase the violence and frequency of storms.

In the past seventy thousand years, natural changes in global temperature have led to rapid changes in the normal ocean currents, including a change in the warm water flowing north out of the Caribbean Gulf Stream such that it no longer flowed northeast to the coasts of Ireland and Norway and then west to Greenland, but instead sunk into the ocean in the mid-Atlantic. The cold Canadian arctic air that blows east is normally warmed as it passes over the northern Gulf Stream. When the warm water path changed, the air reached Europe and created a small ice age.[51]

These climate changes occurred from natural perturbations beyond human control, like changing the tilt of the earth or fluctuations of the sun's output. Today, we may be inducing similar patterns with human-released carbon dioxide.

WE HAVE GROWING ENVIRONMENTAL PROBLEMS

There are several key facts that are beyond dispute. The world population has quadrupled in the last one hundred years. The planet cannot absorb four times more people in one hundred years without harm. To counter the harm we must apply collective knowledge and leadership to improve the standard of living for all people, but also to lessen humankind's impact on the environment.

A second indisputable fact is the increase in worldwide carbon dioxide emissions. The human-induced emissions of CO_2 were trivial in 1860. These emissions rose to 6.4 billion metric tons per year by 1997.

This is 1.4 billion metric tons per year more than was emitted just ten years ago in 1987. (See page 16.)

A third fact is that the current level of carbon dioxide concentration in the global atmosphere is 35 percent higher than at any time in the past 160,000 years and is nearly double the concentration level of fifteen hundred years ago, as can be seen on the Vostock ice-core data (shown on page 19).

Finally, the fourth fact is that the global temperatures have correlated very closely with the concentration of CO_2 in the atmosphere over the past 160,000 years, as measured by Antarctic ice-core samples. (See page 19.)

We observe other relevant facts including: (1) the warmest 11 years on record have occurred since 1980, (2) there has been an increasing frequency of more extreme weather events, and (3) the polar ice caps are melting and thinning.[52] These observations are consistent with the theory of global warming from greenhouse gases, but they could be the result of natural cycles. These data fall short of conclusive proof of human-induced global warming, but evidence keeps growing that human actions are having a discernible effect on climate.

The killing of trees and entire forests by acid rain and other power-plant emissions is also beyond dispute. The entire forest system is suffering. Vermont scientists Richard M. Klein and Timothy D. Reikins find compelling the correlation of pollution products appearing in forest soils in Europe and America, and forest decline in the same areas. These scientists speculate that the increase in pollutants from power plants is interfering with the trees' ability to obtain needed nutrients, and they put forward several possible mechanisms that may link pollution and dying forests. They wrote,

> Tree ring analysis and other studies suggest that the initiation of forest decline in the northeastern United States, adjacent Canada and Central Europe all date to about 1950–1960. While making correlations is a dangerous game, this was the same period when alkaline fly ash was eliminated, when construction of tall stacks on power plants allowed pollution to be widely dispersed ... (when there were) increases in nitrogen oxide emissions. . . .[53]

Charles E. Little, an environmentalist and author, states that,

All over America, our trees are dying. From the "sugarbush" of Vermont and the dogwoods of Maryland's Catoctin Mountains to the forests of the "hollows" in Appalachia, the oaks and aspens of northern Michigan and the mountainsides and deserts of the West, a whole range of human-caused maladies—fatal ozone ultraviolet rays, acid rain . . . has brought tree deaths and forest decline in its wake.[54]

In short, the excessively rapid consumption of fossil fuel in old power plants with out-of-date pollution control is causing local and global environmental problems.

FEAR OF ECONOMIC DISRUPTION COLORS POLICY

The growing list of environmental problems and facts is a very strong call to action. We can do better. Most people would easily reach the same conclusion but for one huge mistaken assumption. Many people, possibly including the majority of the United States senators who will be asked to ratify the Kyoto Protocol on global warming, mistakenly assume that our conversion of the energy content in fuel to electricity or heat is optimal. They assume that to reduce carbon dioxide pollution we will have to use less energy and thus lower our standard of living. They assume that CO_2 reduction will require expensive pollution control and disrupt our economy. These assumptions are not correct.

We can eliminate barriers to efficiency, lead the world toward more efficient use of energy, and save money while lowering pollution. The task of persuading people of these truths is daunting, as spelled out in the *Washington Post* in an editorial:

> A funny thing happened on the way to the international climate meeting in Kyoto, Japan in December. Before the United States could take its place in the negotiations, the Senate last summer unanimously passed a resolution, sponsored by Robert Byrd (D-WV) and Chuck Hagel (R-NE), stating that the United States should not be a signatory to any agreement limiting greenhouse gas emissions unless such a treaty mandated specific commitments for developing nations.
>
> Fossil-fuel industry lobbyists are especially enthusiastic about the Byrd-Hagel language. They now argue that the United States should

abdicate its leading role in reducing greenhouse gas emissions until
President Ernesto Zedillo agrees to clean the air in Mexico City or
Prime Minister Jiang Zemin decides to make the Chinese less reliant on
dirty coal for energy. To do otherwise, they say, will put America at a
competitive disadvantage.

Taking the language of the Byrd-Hagel resolution too literally
would require the United States to forfeit global environmental leadership in a world practically begging for it. . . .

We need to settle on a generous definition of developing countries' "participation" in the campaign to control climate change that will
satisfy the environmentally conscious supporters by Byrd-Hagel
without compromising the United States' ability to act in its own interests. Those countries will catch up if the United States leads the global
race to use fossil fuels more efficiently.[55]

The Clinton administration released an electric restructuring policy in
April 1998 (later introduced as a bill)[56] that would require all states to
deregulate electric generation by 2003 and set guidelines for recovering
stranded costs. If passed, the bill will help reduce greenhouse gas emissions
and improve the economy. However, the administration passed up a golden
opportunity to lead our heat and power industry to greater efficiency. In
the debates leading up to the release of the administration's proposals, misguided environmental activists worried that deregulation would lead to
more pollution (the opposite of what is occurring in the newly deregulated
United Kingdom and Argentina). Facing the political reality of a Senate that
does not understand the wastes of monopoly electric generation, the
Clinton/Gore team offered a very modest proposal for electric restructuring. Restructuring and lessening the monopoly will help the environment,
but we need leadership to guide us toward a sustainable future.

Most of the world's nations convened in Kyoto, Japan, in December
1997 to negotiate a worldwide agreement to curtail greenhouse gas
emissions. The final protocol called for industrialized nations to reduce
greenhouse gas emissions by, on average, 5.2 percent, compared to 1990
levels, with individual countries required to make greater or lesser
reductions. The protocol establishes several innovative mechanisms
which will aid countries in meeting their reduction targets. These mechanisms include allowances for international emission credit trading and
the establishment of a program to transfer sustainable technologies to
developing countries in exchange for emission credits.

On March 16, the Kyoto Protocol on climate change was opened for signature and Antigua and Barbuda, Argentina, Maldives, Western Samoa, Switzerland, and St. Lucia were the first to sign.[57] With 4 percent of the world's population and 25 percent of the world carbon dioxide emissions, U.S. leadership is sorely needed.

The coal, oil, and gas industries have generously funded an antiglobal warming group, the Global Climate Coalition, to lobby against the Kyoto Treaty. The coalition's March 13, 1998 news release stated, "Under the terms of the agreement, the U.S. would have to cut its energy use by 30 percent." This is only true if we continue to protect monopoly electric generation and sales and continue to waste two-thirds of the fuel we burn. The U.S. can cut its fuel use by more than 30 percent and still use the same amount of energy. We must not confuse energy use with the fuel we burn. The oil industry was not unanimous in opposing actions to stem global warming. British Petroleum acknowledged that the evidence for global warming merited attention and study. Royal Dutch Shell originally helped fund the antiglobal warming effort, but subsequently dropped out, and has since announced major investments in renewable energy, especially photovoltaics and biomass.[58]

We can draw several conclusions from the items above. (1) We must tell our elected officials that we care about the environment and the future of the planet, that we fear the consequences of rolling dice with global forces, and that we want environmental leadership. (2) The environmental movement needs to study the wonderful pollution-control opportunities found in increased efficiency of power generation. Environmental regulations that block efficiency are part of the problem. (3) We must match our progress in science with progress in ethics.[59]

PROGRESS IN ETHICS?

Freeman Dyson, British-born U.S. physicist and author, in his latest book, *Imagined Worlds,* is concerned with ethics and how we might mitigate the evil consequences of technology. He attributes to philosopher J. B. S. Haldane the idea that progress in science will bring enormous confusion and misery to humankind unless it is accompanied by progress in ethics. With respect to global warming and acid rain destroying our trees, there is an ethical issue. We are burning too much fuel in inefficient power

plants. Then we waste much of the energy that was produced. Is this ethical? Haldane contends that science has given us "much" and the "much"
that is required involves ethical stewardship of our resources.

It is hard to see how the human race will rapidly develop a deeper
sense of ethics. We should try, but controlling the power we now have
to liberate the energy in fossil fuel and changing the global climate and
local environment is today's problem. Somehow, we must find a way to
increase human thinking power. We cannot sit back and wait for ethics
to develop to a higher level. Self-interest is a fact of life.

A PROVEN WAY TO INCREASE
HUMAN THINKING POWER

We must face the corollary of having god-like powers and god-like
responsibilities for the future. If we do not recognize this responsibility,
and if we cannot increase our thinking power geometrically to exercise
this responsibility, we may do irreversible damage to our Earth.

Yet, no one, whether a Fortune 500 executive, a craftsman, a
working mom, a Nobel laureate, or the president of the United States,
has the remote possibility of exercising god-like responsibility for the
future. What we need is a collective, global cooperation to search for the
knowledge to move forward responsibly. We must take many local
actions that are based on global consequences to increase the probabil-
ities of our collective survival, and this requires a lot of thinking—too
much for any single individual. How do we harness our collective thinking
power? How do we keep local parochial interests from producing environmentalist Garrett Hardin's "Tragedy of the Commons," where pursuit
of individual self-interest results in collective disaster?[60] Is there a way?

Parallel processing is an approach to handling complex problems. For
years, scientists and engineers have tried to make faster and faster computer-processing units so they could model some of the really big problems such as global weather. The weather models are complex, and they
take vast amounts of computer time on even the fastest known super-
computer. In the past ten years, there have been great developments in
parallel processing, where problems are divided into subparts and many
computer-processing units each take a piece of the puzzle. Sandia Labs,
one of our national laboratories in Albuquerque, New Mexico, has just

commissioned a computer that uses 9,200 Intel Pentium processors that operate in parallel. These 9,200 processors are not unlike the one chip driving the computer helping me prepare this chapter, but they are interconnected and have the capability of analyzing, in a short time, problems that were previously untouchable with the biggest, fastest super computers. Parallel processing provides a model for increasing the power of human thought.

We need a way for humankind to act as a giant parallel processor, testing what works, retesting every theory, acting to quickly correct the problems we create, spreading the successful actions rapidly. In short, we must work as a collective thinking organism. Although this may sound far-fetched, it is not.

We have already developed and perfected a method of increasing human thinking power geometrically, of connecting all human brains and actions. This method is one of the great social inventions of human history. This invention also provides a proven mechanism that causes group ethics to be an improvement over most individual's ethics. It self-corrects and is amenable to leadership. This invention evolved in antiquity, and only in the past two centuries did thinkers begin to recognize the magical power of this social invention to promote the general welfare. In *The Wealth of Nations*, first published in 1776, Adam Smith was the first to clearly recognize the invisible hand of the marketplace.[61]

Smith explained how each individual, acting in his or her own self-interest, nonetheless would advance the general welfare. He showed how the self-interest of each individual, if unfettered by restrictions, would lead to advances in every process, in every field. He helped us to understand how competition harnesses self-interest to promote the general welfare. We call this marvelous social invention a free market, and what drives markets to improve the general welfare is, paradoxically, human self-interest. Self-interest can be good for society, provided it is not sheltered from competition.

Market actions are unpredictable. Government legislation that assumes future technology and mandates solutions to solve current problems invariably turns sour. Governments often prevent competition from acting properly by restricting competition in various fields. The defense of these government restrictions is that competition will be wasteful; it will be better to give monopoly powers to a government-owned business, or to selected private organizations controlled by government. The

1974 Nobel laureate and Austrian-born economist, Fredrich A. Von Hayek (1899–1992), was deeply impressed by the power of markets. He observed that roughly thirty thousand years ago, human affairs became too complicated and intertwined for anyone to predict the market actions. The best legislation sets goals and guides society, but leaves details to markets. Government can rethink its approach to regulation. It can learn from Von Hayek.[62]

The market, as described by Adam Smith and a long line of economists down to Von Hayek, works only when true competition is allowed. Whenever governments restrict competition, markets do not work. When we legislatively enshrine the perceived truths of one age, we lose the self-correcting beauty of markets. We lock in inefficiency. We become the Grand Inquisitor and Galileo is our prey.

After fifty years of central planning by the Soviet Union, the entire economic system collapsed. The world is too complicated to manage effectively without market forces continually correcting problems and developing new solutions. We need interconnected thinking of all human beings with rapid error correction to manage the complexity. We need to allow this error rapid and continuing correction to work to achieve efficient solutions that mirror the latest knowledge and technology. However, there are several problems with an absolutely uncontrolled market. Markets work when there are at least several strong competitors who police each other, pointing out any problems with the other competitors' approaches, offering lower prices and better quality products. To ensure that the market exists, i.e., has several competitors, governments need to prevent one firm from monopolizing a particular market.

For purposes of this book, a second problem of unfettered markets must be addressed. Functioning markets will find the most efficient solutions, but those solutions will not take account of externalities, of impacts on society that don't affect the cost of production. The classic example is damage to the environment caused by production. This damage hurts the population but does not, without government intervention, represent a cost to the producer of the good or service. The second essential role of government is to establish and enforce rules that either limit what anyone can do to the environment, or levy fees for pollution. Markets then find the optimum solutions that fit within the rules and imposed costs.

Government's important tasks are to assure that competition is open, and to set and administer rules for all competitors that preserve the environment and demand ethical behavior. Fortunately, this is pretty much what governments do in most areas of the economy—except for energy. Government's job is to assure the functioning of the human parallel-processing system. The United States was founded on these philosophies and has proven the worth of free capitalism to the world. By harnessing self-interest within an ethical framework, a long line of U.S. administrations have set in motion the geometric growth of knowledge and actions that are vital to managing our growing god-like power over nature. Government must put in place the rules that protect the planet and its many life forms, but must not attempt to choose the detailed solutions. Government's job is to set the goals, at the highest possible level of generality. It must then get out of the way and let collective self-interest, expressed through market forces, achieve the goals and promote the general welfare.

Our growing climate change problem is occurring precisely because governments have shackled markets in the energy sector. With electricity shielded by a monopoly cocoon (and until recently natural gas as well), energy decisions have been subject to government regulation instead of competition. It is a bad trade-off. In the next chapter, we will explore how enlightened policy with respect to energy can unleash market forces that will reduce the human-related carbon dioxide output. These enlightened policies will speed us toward a sustainable energy future and at the same time improve the standard of living of today's citizens.

4

Sending Price Signals

UNLEASHING MARKET FORCES

On September 26, 1997, I attended a gathering of senior representatives of fourteen industries at the White House, chaired by White House special assistant Todd Stern. It was one of a series of meetings to gather information on how the United States should approach the Kyoto negotiations on climate change. Administration staff included Dr. Janet Yellen, chairman of the Council of Economic Advisors; Katherine McGinty, chairman of the Council on Environmental Quality; and Dirk Forrister, chairman of the White House Task Force on Climate Change.

Todd Stern indicated that President Clinton had asked for intense policy debate to arrive at appropriate policies, and that the debate was to be guided by four principals, including:

- The science about climate change is clear enough to warrant action.

- The president is committed to endorsing binding targets for carbon dioxide reduction.

84

- The president prefers as much flexibility as possible.

- Developing countries need to participate in any solutions.

Industry Is Energy-Conscious

As various industry representatives spoke, a fascinating pattern emerged. Where market forces were at work, companies had paid considerable attention to energy consumption and cost, and had built impressive records of reducing energy-use per unit of product. We each had aggressive goals for further energy use reductions per unit of product. Paul Cicio, Global Issues Manager of the Dow Chemical Company, told us that Dow Chemical had reduced its worldwide energy use per unit of product steadily over the past ten years, and had goals of further reducing the energy use per unit of product by one percent per year for the next ten years.

Dow cogenerates a phenomenal 95 percent of its electricity worldwide. Cogeneration, or combined heat and power, is critical to this discussion because it has been Dow's most important way to reduce carbon dioxide and save money. It involves generating electricity at Dow's manufacturing facilities where their processes or buildings use heat, then capturing the normally wasted heat to feed those processes. As explained in chapter 2, only 33 percent of the energy in the fuel is converted to electricity on average throughout the United States, and the remaining 67 percent of the fuel's energy is converted to heat, which is usually thrown away. By generating electricity at its manufacturing plants, Dow can recover the heat and use it in various processes to manufacture chemicals. Cogeneration is two to three times as efficient as separate generation of electricity in one plant and heat in another. *Most important, Cicio conceded that Dow pursued these energy reductions out of pure self-interest.* If Dow does not keep reducing product cost, they lose market share. Competitive market pressure causes Dow to conserve energy.

Representatives of Bethlehem Steel, Dupont, Georgia Pacific, and Weyerhaeuser echoed the same story. Each manufactures products which are sold in large quantities to sophisticated buyers; every nickel of cost matters. Mike Thompson echoed the story with respect to Whirlpool's production costs for appliances. His company and industry has dramatically reduced the energy used to make each product and the energy used by the product. They are focused on further energy reductions. For example, in 1972 the industry-average refrigerator used 1,726-kilowatt hours per year, while the 1995 model uses 649-kilowatt hours per year to do the same job—a 62 percent energy reduction. Comparable reductions from other appliances are freezer, minus 68 percent; room air conditioner, minus 35 percent; clothes washer, minus 42 percent; and dishwasher, minus 50 percent.[63] However, Thompson introduced us to a major problem—the disconnect between energy-efficient products and consumer demand.

Consumers Are Not Energy-Conscious

Figure 11 shows the lifetime energy use of a clothes washer. The production, distribution, and ultimate disposal of the clothes washer account for a tiny fraction of the energy consumption and associated air pollution. Over 95 percent of energy consumption and air pollution come as a result of the washer's use. Other appliances have similar patterns. Thompson said that to persuade consumers to demand energy-efficient appliances, we must have federal appliance-efficiency standards. Why would a businessman ask for federal efficiency standards for his products?

According to Thompson, the typical appliance customer spends only about five minutes with a salesperson. The customer asks first about features, then price, and then availability of the model desired. Can it be delivered? Will the delivery person remove the old clothes washer? Then, once in a great while, a customer will ask about the energy efficiency of the washer versus other brands. Seldom does one focus on the energy use of the appliance over its life, even though it might be a terrific investment to pay a few more dollars for a machine that uses less energy every time it washes a load of clothes. The market is not working perfectly. Although it would make sense for consumers to pay more up front for an efficient model that would save energy and cost on every load of clothes washed over the fifteen-year life of the machine, there is little demand for energy efficiency. Why aren't consumers saving money by investing in energy-efficient machines?

Thompson's revelation led me to do some research on what was available in energy-efficient clothes washers and dryers. A search of the Internet quickly turned up a Department of Energy web site that rates available clothes washers for energy efficiency and lifetime efficiencies. A German-made Miele, horizontal drum, front-loading washer uses only 20

Fig. 11. Cradle-to-Grave Energy Consumption of a Clothes Washer[64]

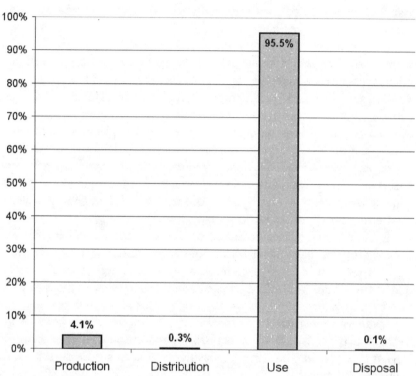

Energy Consumption

percent of the energy of a standard American top-loading washer. The horizontal drum does not need to immerse the clothes in water since they tumble constantly, so it uses less than half as much water per load of clothes. This saves hot water and detergent. Since the horizontal drum rotates in one direction and tumbles, far less energy is expended in the Miele's wash cycle than a vertical drum machine that agitates clothes by moving left and right. Finally, the Miele spins the clothes at 1600 rpm versus 600 rpm in conventional washers, removing 35 percent more water with increased centrifugal force. This is a far cheaper way to remove water than using gas or electric heat to evaporate the water in a dryer.

The Casten household needed a new washer, so this seemed a good way to do "mother-in-law" research. The efficient Miele or Adzo (Swedish) models are available locally, but are not tailored to the U.S. market. Physically smaller than their U.S. counterparts, they cost $1200 to $1500 versus $500 for a bare-bones American top loader. The requisite 220-volt electrical service may require a special electrical installation that costs up to $1,500 more.

U.S. manufacturers could develop a machine with similar energy demand and sell it at a lower price *if there was an American market.* This reflects a failure of education on values. U.S. consumers are not aware of the damage their profligate energy use does to the global environment and the United States economy, so they do not, in general, demand or even investigate efficiency, even though efficient appliances could represent a guaranteed 20 percent return on the extra investment. American consumers have so much wealth and disposable income relative to most of the world that they often can't be bothered with using less energy, even when it saves them money. This lack of focus on product energy efficiency can be demonstrated across the board for U.S. consumers.

John D. Hopkins, vice president of Government and Public Affairs for Owens Corning, had a similar story of market disconnect. Owens Corning uses one unit of energy to make a piece of fiberglass insulation, and that same insulation saves 12 units of energy every year of its life—which may be fifty to one hundred years. Many homes and buildings could earn high returns on investment by increasing the amount of insulation. There is an industrial market for energy-saving insulation, but the consumer market lags behind. The net present energy cost of a house—that is, the cost to build and all the future costs to heat and cool over fifty to one hundred years, all discounted from the time the costs occur to the present—would be lower if more insulation were used in construction.[65] Consumers demanding houses with better insulation would save money and energy. Why do we fail to optimize energy choices? Education? Values?

Everyone who works in a commercial office building sees another example of market breakdown with respect to energy use. Trigen Energy Corporation's mission is to produce heating, cooling, and electricity with half of the fossil fuel and half of the pollution of conventional generation. My colleagues are fanatics about energy efficiency; they constantly focus their considerable intellects on reducing the fuel burned in our power plants. They track the energy and pollution per unit of production, compare both to conventional energy use and pollution, and then reward and recognize people who improve either variable. Of all firms, Trigen should be aware of energy use. It routinely performs the sophisticated analysis needed to determine the payback of investments to save energy. Sounds good, but look at our leased corporate offices in White Plains, New York.

Lights turned on in our conference rooms for meetings are seldom turned off when meetings are over. Computer terminals are left on twenty-four hours per day, unless they are the new energy-saver models promoted by the DOE that switch to low power after a period of inactivity. Even then, most terminals are not set to turn off, because the users don't like having to move the mouse to cause the screen to turn back on. Almost no one turns the lights off in his/her own office when they leave for lunch or for the evening. The cleaning staff comes in at 5 P.M., goes through each office to pick up the trash and recyclable papers, and turns on every light. These lights are not turned off again until 8 to 10 P.M. when the vacuuming and dusting is complete. All the lights burn needlessly in every single Trigen office for three to five hours in order for the cleaning people to have three to five minutes of illumination in each office. Lights for the common areas including halls and restrooms are never turned off.

In the warm months, the energy use problem is compounded. The needless electric use gives off heat, which must be removed, so the air-conditioning system is forced to consume more energy to remove the heat from the wasted light and unattended computer terminals. Yet all of this is correctable. Newly developed occupancy sensors detect movement and infrared radiation emitted by our warm bodies. They automatically turn off the lights in unoccupied rooms. Occupancy sensors are little used in America even though they save energy and money. Why doesn't the market work?

Amory and Hunter Lovins and their colleagues at the Rocky Mountain Institute have analyzed how much energy could be saved by eliminating inefficiency and needless energy use, and by installing more efficient lighting and fans. Amory coined the term *Negawatts* to capture his vision of how we could meet rising demand for comfort and lighting. He demonstrated that the use of higher-efficiency lighting, more efficient systems for cooling and ventilating buildings, and wide use of occupancy sensors to turn off lights in empty rooms would reduce the electric demand, or produce negawatts. This approach would meet our needs at a significantly lower cost than is involved in building and operating extra electric generating capacity to supply the wasted energy. There has been a considerable push from regulators to foster the installation of these devices, commonly referred to as *Demand Side Management* (DSM). Some progress has been made. One thousand and three of the 3,199 electric utilities in the United States reported in 1996 having DSM programs with claimed energy savings from these programs representing 2.0 percent of annual electric sales—a small token of what is possible. The total expenditures on DSM by utilities in 1996 were only $1.9 billion, under 1 percent of total electric revenues.[66] The overall deployment of proven technologies to reduce energy consumption and save money is quite small. Why?

WHY AREN'T ENERGY-SAVING TECHNOLOGIES DEPLOYED MORE RAPIDLY?

Dr. Janet Yellen, chair of the Council of Economic Advisors (CEA), raised a most interesting question. After hearing the consumer-product company representatives tell about the technologies they had developed that would save energy, she asked, "Why aren't these technologies being deployed more rapidly?" I think this critical question was appropriately raised by the president's chief economist. At root, this is an economic question.

Markets Depend on More than Price

For Ludwig Von Mises, the renowned Austrian economist, the proper study of economics was human action, so much so that his opus is titled *Human Action*. According to Von Mises, classical economists have failed to realize that the theory of value

> was much more than the substitution of a more satisfactory theory of market exchange for a less satisfactory one. In making his choice man chooses not only between various material things and services. All human values are offered for option. All ends and all means, both material and ideal issues, the sublime and the base, the noble and the ignoble, are ranged in a single row and subjected to a decision which picks out one thing and sets aside another. Nothing that men aim at or want to avoid remains outside of this arrangement into a unique scale of gradation and preference.[67]

In other words, the choices each person makes encompass more than simply profit motive, but also include a host of other variables and considerations. Why does the typical commercial landlord all over America not install devices that reduce his/her energy costs, even when those devices represent excellent investments? We learn the answer from Von Mises's view of human action.

- Besides price and return on investment, the landlord's decision factors in the organizational competence and how much effort is involved to learn about the technology relative to total bottom line impact.

- Has the approach a good reputation? Gary Fechter, a colleague of mine, was a physics instructor at West Point fifteen years ago when the military academy installed the earliest version of occupancy sensors in classrooms. The sensors would turn out the lights in the middle of Gary's lectures, at which point the cadets would all clap, causing the sensor to turn the lights back on. This was the original "clapper," predating today's Clapper™ that is designed to turn lights on or off with a handclap. These early failures gave occupancy sensors a bad reputation that still endures in spite of near-perfect technology today.

- Is the practice common and acceptable? Most people prefer not to be pioneers.

- Do the landlord's customers care about energy efficiency? Would increasing energy efficiency help to lease space?

- Would a malfunction annoy tenants and destroy good will?

Until most of the answers are positive, Americans will continue to needlessly fill the global atmosphere with carbon dioxide. Limited to pure profit analysis, the failure of landlords to invest $100 in an occupancy sensor in every restroom, each of which would reduce electric use by $40 per year, or yield a return of nearly 40 percent interest on the investment, appears to be a failure of the market. This assumes the only inputs to a landlord's decisions are capital investment and potential savings. In Von Mises's view, "*all ends and means, both material and ideal issues...*" go to explain why a decision "*picks out one thing and sets aside another."* The landlord does not want to expend the time and master the complexity of high-tech energy-saving devices just to save a few dollars. He or she can make more money by focusing on real estate. In a fully free market, entrepreneurs spot these opportunities and create value by mastering sophisticated methods that are cheaper than the common practices. In a fully free market, energy companies fighting for market share would focus on conversion of the energy content of fuel to electricity and heat and the efficient use of that electricity and heat, and would take responsibility for energy decisions in return for a profit.

Energy Efficiency Is Proven High Return and Low Risk

At the New York Earth Day Conference on April 22 and 23, 1998, Kathleen Hogan, director of Atmospheric Pollution Prevention Division, U.S. Environmental Protection Agency (EPA), unveiled a new joint program with the Department of Energy (DOE) to award an "Energy Star" designation to commercial buildings that are in the top 25 percent of all buildings in energy efficiency. This designation will signify that the building has utilized energy-saving technologies like insulated or special reflective glass, on-site power generation with heat recovery, photo voltaic panels that produce electricity from sunlight, solar water heaters, occupancy sensors, and other technologies to reduce its energy waste. The objective is to recognize responsible builders and real-estate developers.

Hogan made this startling statement, "Commercial buildings in the United States waste 30 percent of the energy they use, and this costs $25 billion per year." How can commercial buildings, which are subject to market forces, waste $25 billion per year? Is the necessary investment too high? Figure 12 uses information from the EPA and Vanguard, one of the country's largest investment managers, and it shows that efficiency is a very good investment.

Investors relate risk and reward. The higher the volatility or risk of a given investment, the higher the reward or return that is demanded by investors. What Vanguard says in figure 12 is that U.S. Treasury Bills, with a very low volatility, are bid down to an average annual return of 5 percent, while highly volatile or risky small company stocks must pay returns over 20 percent per year to attract investors. The EPA's analysis is that energy-efficiency investments have a low risk comparable to corporate bonds. This makes sense. As long as the building is occupied, comfort and

Fig. 12. Building Energy-Efficiency Risk and Reward
Versus Other Investments[68]

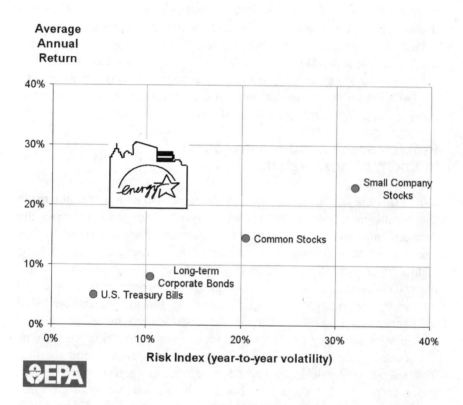

Average
Annual
Return

Risk Index (year-to-year volatility)

lighting will be required, so efficiency technologies save energy every year, and the amount of money saved varies only with the price of energy. According to the EPA, these technologies, at today's prices, produce 20 to 30 percent returns on investment. This is the type of investment that will always be desired—high return and low risk.

Why then are commercial buildings wasting $25 billion of energy per year? Remember Von Mises's analysis of human action. There are many factors besides price involved in landlords' decisions to master and invest in energy-saving technology or to continue wasting energy. Failure to invest in efficiency is not simply an energy price issue. If landlords or the energy companies who supply buildings do not make every low-risk, 20 to 30 percent return investment possible, would they suddenly move to reduce energy use if the price of energy rose 10 to 20 percent? There are other factors at work which governments can correct. There is a market failure which goes beyond price and investment return. How will these factors be corrected by fuel taxes alone? The Economist recently published an editorial advocating carbon taxes, stating, "... the easiest way to reduce something is to tax it—in this case taxing the carbon content of power."[69] I disagree. The easiest way to reduce waste is to unleash market forces. Only after competition is allowed can we see if we need taxes to further discourage use of energy.

Problems with Landlord Investment in Energy Conservation

Why does the landlord, out of pure desire to increase profit, not make the installation of energy-saving technology? Besides the complexity the landlord must master to make the investment, there are several other reasons, all of which must be addressed for a market to function to reduce energy consumption. To begin with, the utility costs in a commercial office building are nearly always passed through to tenants, based on the square footage occupied. Commercial tenants are seldom billed for energy based on metered usage. We exclude here the building owned and fully occupied by one user, but focus on the typical office building that houses various businesses, accounting firms, lawyers, and the like. The spaces are periodically redivided as tenants require more or less space or move in or out of the building. Large buildings have a permanent crew responsible to reconfigure spaces as occupancy changes. It is a physical problem to meter the electric or the thermal usage of individual spaces. To do so would require either connecting all of the wires to a single source each time the space is changed or using multiple meters. (Advances in meters make measuring an individual tenant's energy use much less costly than in the past, but the regulations are all based on decades-old analysis.) All tenants must share pro rata the energy costs of the common spaces like halls, elevators, restrooms, and entry foyers. The heating and cooling usually is produced in a central boiler and chiller plant, or comes to the building wall from a district energy company, and is for the entire building.

It is feasible but difficult to measure the heat or cooling used in the space each firm occupies in a communal office building. Instead of measuring energy use, the typical landlord simply adds to the rent a utility charge per square foot of space occupied. A typical surcharge in New York City is $2.00 to $2.50 per square foot per year. Once the utility charges are divided pro rata over all of the rented space, there is no incentive for the individual renter to conserve energy, and the only value to the landlord for conserving energy is to have a lower utility charge than competitive buildings. Given a base rent of, say, $25 per square foot, a reduction for energy of $.25 to $.75 per square foot is not worth the landlord's time and effort to master energy-saving technology. With increasing environmental awareness, some landlords have acted to make their buildings "green," and use this as a marketing tool to attract tenants. In this context, a building is "green" to the degree it employs electricity- and thermal-energy-conserving technology and/or uses renewable energy to produce the needed electricity and thermal energy. The practice remains the exception rather than the rule, however, and is applied primarily to new buildings.

This is a market disconnect. If one tenant assiduously turns off lights and carefully reduces energy consumption, the landlord profits. He or she is unlikely to pass through the savings to all tenants, and even if the landlord did share the savings, the reduced utility charges would be spread to all other tenants. This is an example of Garret Hardin's "Tragedy of the Commons."[70] The actions that are in the best interest of the individual are often not good actions for the society. Hardin's title refers to the English custom of having a grassy *Common* in the center of each town where residents may graze their cows or sheep. Each individual gains by grazing several animals, but if all individuals each graze several animals, the common is destroyed, and the animals all go hungry. If conserving energy in an office does not produce rewards for the individual, the individual finds no logic to do what is best for the society and in this case the global environment.

To compound the problem, in many states it is illegal to submeter an individual tenant's energy use. The theory behind these laws is that landlords would resell electricity and make a profit, so to protect tenants against the assumed greed of landlords, submetering was made illegal. In this case, laws and regulations to prevent profit making also prevent market signals. The laws keep the tenants from realizing the cost of using energy. Since the tenant's costs do not change directly because they leave lights burning, the tenants often leave their lights on. The landlord must pay for the wasted electricity, and then raise the rent to cover the cost of the wasted electricity. The tenants pay for their energy waste via higher rent payments, but they do not see a direct connection to turning out the lights. They do not change their behavior and conserve electricity, because of the ban on submetering their electric use. The result of these laws is that we use too much energy.

Asking Utilities to Invest in Conservation— Demand Side Management

Many environmentalists know we waste energy and damage the environment, but don't seem to see that monopoly protection of electric production and sale is the reason the market does not work. Instead of demanding deregulation, they have persuaded public service commissions to force the regulated electric utilities to institute Demand Side Management (DSM) programs. This refers to installing efficiency devices to lower or manage the peak electric load or demand. In theory, as is well documented by Amory Lovins, a physicist, vice president and founder of the Rocky Mountain Institute, it is cheaper to reduce peak electric demand by installing energy-efficient devices than to build new power plants. The regulatory commissions are not typically charged with protecting the environment, and their deliberations on DSM have focused on saving capital costs. The utilities have been encouraged to invest assets and expertise to relamp commercial office buildings by being allowed to earn a regulated return on their incremental investment. I explain below why this is a less than optimal way to encourage energy efficiency. Start with a mistaken view of why the market is not working or even a conclusion that markets will not work, and you must then invent other rules to bring about the desired behavior. The market is not working in commercial office buildings because electric utilities have monopoly protection.

The easiest way to fix a problem caused by monopoly protection is to end the monopoly. Mandatory Demand Side Management seeks to fix the problem without eliminating the cause. This is comparable to attempts by medieval high priests to justify their claim that Earth was the center of the universe. When real-life observations of planetary movements suggested the sun might not revolve around Earth, the Earth-centered proponents invented epicycles. This truly imaginative theory held that the sun certainly circled the earth each day, and then did a few pirouettes—small orbits of its own as well, which would explain the observations. This was a bad patch on a bad theory. Copernicus risked the wrath of the priests by proclaiming the sun, and not the earth, to be the center of the universe. DSM is a similar bad patch on a bad theory.

I have never met a single electric utility executive who believed DSM was good or logical for his monopoly business. These executives have directed their utilities to support modest DSM programs to go along with the regulators in order to keep the peace, to satisfy enough political interests to preserve their monopoly rents. These utilities have the absolute right to be the only supplier of electricity to a building, and they earn profit by selling that electricity. How can it possibly be in a utility's interest to deploy technology that helps buildings use less of their product? Deployment of energy-efficient devices with DSM programs requires the utility to go through cumbersome rate hearings to increase their rates enough to earn an 11 to 12 percent return on the cost of the DSM devices. If they do not succeed in deploying as much energy-efficiency as is warranted, they are allowed to build new generating capacity and earn a return on the investment in that capacity. But if Amory Lovins is right that energy-efficient devices cost less per "negawatt" than the cost of a new megawatt of generation capacity (and I think the analysis is correct), then the utility has a serious motivational problem. The utility profits increase when the utility is allowed to increase its capital investment, i.e., increase its rate base. Monopoly utility shareholders benefit when management opts for the course that has the highest capital cost and thus allows the highest addition to profits. It is better for the utility shareholders to have expensive new megawatts built than to have less expensive "negawatts" generated with efficiency investments.

Instead of rate commissions pushing monopoly utilities to act against their own best interest, consider the motivations if the building owner was legally allowed to choose energy suppliers—if the states stop using police powers to prevent competition. In a free market, energy companies could earn as much or as little profit as the market would allow by using energy efficiency as a tool to win the energy-supply contract for the building. Instead of dreaming up a bad patch on a non-working market, it would make more sense to fix the market. To do otherwise is to encourage hypocrisy and further confuse the energy-efficiency debate. DSM has not been very widely installed relative to its economic potential.

Cultural Values Impact Decisions about Energy Efficiency

Cultural values are also involved in the landlord's failure to install energy-saving devices. Other societies with different cultural values use significantly less energy in offices. I recently visited Japan, and went with my host, Shichiro Matsukata, to his closed office building in Tokyo on a vacation day. When we passed the security guard, the halls were almost dark, illuminated by very low-intensity green light just bright enough to let us find the light switches and elevators. These small safety lights used only one-twentieth of the energy use of full lights. A sensor detected our presence and switched on the lights in the hall. We went up to the sixteenth floor to Shichiro's offices, and turned on the lights in a large room filled with desks. An occupancy sensor measured both movement and infrared radiation typical of human beings. Five minutes after we left, the lights were automatically turned off. The restrooms were all equipped with the same occupancy sensors.

Japan subscribes to the antiquated theory of electricity being a natural monopoly and uses state police powers to prevent competition in electricity, yet the Japanese use less energy per square foot. Part of the reason is price. Japan has little indigenous fossil energy, must import all of its fuel, and has taxed energy to discourage use. This is the classic "carbon tax," and it clearly has an impact. It sends a strong economic signal to all Japanese people to reduce their consumption of energy. The Japanese have a deep national concern about energy use since every yen spent on fuel leaves Japan. Small schoolchildren learn that it is patriotic to save energy, that their nation has limited fuel resources, and that energy conservation is good. The results of such education profoundly influence Japanese behavior.

At the Matsukata home, we were introduced to the typical Japanese bathroom. A small, deep tub is filled with hot water where one can blissfully sit and soak. The tub is not for washing! Each family member enters the bathroom in turn, then first sits on a small stool on the tile floor next to the tub, uses one small bucket of warm water to wet themselves, soaps down completely, and uses another bucket of warm water to rinse off the soap. Now sparkling clean, they enter the hot tub for a luxurious soak. On completion of the bath, the tub is covered with insulating panels and left ready for the next family member. The small size of the tub and the reuse of the hot water limit energy consumption. Bathing in a communal tub reduces energy from daily bathing by about 70 to 80 percent versus American practices and the results are enormously pleasurable.

In the terms of Ludwig Von Mises, the Japanese actions are not only conditioned by the profit motive, but also informed by a host of other values. The result is a very energy-efficient way to bathe in comfort. The practice saves more money than it would in the United States, due to higher energy prices, so there is more economic incentive, but much more is at work. Widely diffused, the communal bath is supported by Japanese manufacturers and homebuilders. Building such a bathroom in the United States would require some interesting consultations with your friendly builder and plumber, and you would probably have to import the special tub from Japan yourself. The practice makes the Japanese feel good—feel like they are doing the right thing. They are reacting to a whole range of stimuli, including price and savings, and acting in a way to conserve energy. The practice demonstrates that one can save energy and money and still enjoy a soak in a hot tub.

Another Japanese innovation to encourage recycling of aluminum cans sheds light on how to affect cultural values, change behavior, and move to sustainability. At Nagoya Castle, in Nagoya, Japan, there is a recycling machine next to the soda-vending machine. But instead of giving a few Yen back for the can, it is a slot machine. Put a can in, three wheels spin, and if three cherries come up, the machine gives a Nagoya Castle prize. Children (including not so small or young) scurry around the grounds to find a can so they can play. In the process, a cultural value toward recycling is created and strengthened.

How then does America move to an overall set of stimuli that will lead the general public to practice energy conservation? Presidents and their cabinets have to various degrees exhorted the public to conserve energy, as did President Carter, as a result of the two OPEC (Organization of Petroleum Exporting Countries) price shocks. Some practices were changed; growth of fuel use slowed for a few years, but the lasting impact was small. Because of the vastness of America and the abundance of natural resources, Americans have always been fairly uninterested in saving energy.

When Rudolph Diesel invented the diesel engine in the late nineteenth century, it was significantly more fuel-efficient than the Otto cycle engines. (Spark-ignited engines like the typical lawn-mower engine employ the Otto cycle.) Here is the story from a recent history of Cummins Engine Company.

> His [Diesel's] plan was to make money through royalties—that is, by licensing manufacturing rights to leading firms on several continents. Between 1892 and mid-1897, he awarded licenses in Germany, France, Sweden, Denmark, Switzerland, Great Britain, Belgium, Austria-Hungary, Russia, and Egypt. Conspicuously absent from the list was the United States.[71]

All of the Europeans saw the value of a more efficient engine and each wanted their country to have its own engine builder. Diesel thought that he could strike it rich by selling the rights to build his engine in the United States. Finally he met Adolphus Busch, the founder of Anheuser-Busch and one of the richest men in America. The story continues:

> Busch agreed to pay Diesel one million marks (about $238,000) plus a 6 percent royalty for the sole U.S. and Canadian rights to the engine ... he organized the Diesel Motor Company of America in New York, capitalized at $1 million on January 4, 1898. Busch ... seemed to possess a winning combination: monopoly rights—in the world's leading industrial market—to an enormously promising technology ... substantial capital reserves and ... access to all diesel developments. ...
> After two corporate reorganizations and millions of dollars in additional investment, he failed to build a viable diesel business by the time his monopoly expired completely in 1912.
> During the period of 1900 to 1919, diesel engines were being built and put in service throughout Europe, but none were built in America. On February 10, 1919, Clessy Lyle Cummins, a mechanic-inventor and chauffeur, launched a new business venture, backed by his boss, another entrepreneur named William Glanton Irwin in Columbus, Indiana. Mr. Irwin provided most of the $50,000 of starting capital for Cummins who, "proposed to build diesels, a type of internal combustion engine that was still rare in the industrialized world, more so in America than in parts of Europe.[72]

This led to Cummins Engine Company, which as late as 1940 owed such a huge backlog of interest and stock dividends to the Irwin family that if the family ever demanded payment in full the company would be bankrupt. The surge of orders for the war effort finally put the company on solid footing.[73]

This story has always struck me as illustrative of American attitudes toward resources in general and energy resources in particular. We grow up in a land of plenty and assume that the country will never run out of fuel or natural resources. By contrast, Europeans and Japanese grow up in countries with very few natural resources and little indigenous fossil fuel. For them most of their fossil energy must be imported, impacting their country's balance of payments. Consequently, their governments, by imposing high taxes on fuel, discourage wasteful use, and this affects the way the average citizen looks at energy use. Finally, they have all fought repeated wars that were over the rights to energy and natural resources, and often these wars were lost because the war machines ran out of fuel. Today, the resource that is running out is the global atmosphere's ability to absorb all of the carbon dioxide we emit by burning excessive fossil fuel. The availability or nonavailability of fossil fuel in each country can no longer be the only driver of attitudes toward energy conservation.

CHANGING AMERICAN VALUES

Educating the public to believe that conserving energy is the right thing to do is a huge challenge and will involve every known process of altering and inculcating values. How do we change the whole set of "material and ideal" notions in the consumer's head so that energy conservation becomes a central value? Businesses can offer energy-saving products, manage the complexity of energy-saving investments, and advertise new, more efficient products. Changing values to respect conservation will cause energy use to drop and consumers to save money. *But the value changes must come first.* Recent experience with recycling proves that the values can be changed in a relatively short period of time. The separation of household refuse into recyclable and nonrecyclable portions and curbside pickup of recyclable material is barely thirty years old anywhere in the United States, yet has become nearly ubiquitous. Landfill volumes are significantly lower today than ten years ago, and much material is being reused. In 1960, consumers recovered, for recycling and composting, only 62 pounds per year of their solid waste. By 1994, with education and government leadership, the average consumer was recycling 380 pounds per year, an increase of 613 percent. With a population of 260 million, the country recycled or composted 49.5 million tons of material in 1994.[74]

Whole industries have sprung up to convert recycled materials into new products. It has become so fashionable to use recycling that the fast food giant McDonald's abandoned white bags, adopted brown bags made with recycled fiber, and now uses this as a marketing tool. It also saves McDonald's money. Competitors have adopted similar recycled-packaging practices. We can observe a new and important cultural value that reduces the use of energy and trees.

If human action is not just a response to price and savings, but a response to all of the stimuli the consumer faces, how might we address the problem of energy conservation in commercial office buildings with market forces? To begin with, smart energy-conversion devices require a significant amount of sophistication. Their use involves change. The study of diffusion of innovations tells us that diffusion of any changes requires successful trials with observable positive outcomes, followed by peer-group acceptance before the innovations become generalized.[75] Yet, it is extremely difficult for the consumer to observe the savings brought about by investing in more efficient light bulbs when there is no submetering allowed. No observed savings, no changed behavior.

You might think that the landlord who installs energy-saving devices could observe the savings in the building's electric bills. Relamping an entire floor will reduce the next month's electric bill if all other things remain equal. But all things don't remain equal. The weather changes (requiring more or less cooling), affecting the electric use and cost. The fuel-adjustment portion of the electric bill changes, changing the amount paid for each kilowatt. The electric rates vary from season to season, and often base a significant portion of the month's charge on the highest fifteen minutes' peak demand over the last twelve months. This means that the building pays a significant part of its electric bill for the capacity to serve the building, based on the highest fifteen-minute or one-hour demand for kilowatts in the past twelve months. Hence, when the new and more efficient light bulbs are installed, they won't reduce the demand charge for twelve months, by which time there have been other changes in use and prices of electricity. In short, there are no simple ways to observe changes to the total cost of the building's electricity without sophisticated analysis by energy professionals.

The expected savings from installing efficiency devices is subject to calculation and does not even need a trial for people who specialize in energy. Energy firms understand electric rates and the impact on overall consumption from each innovation. These firms can model the savings and make investments with great confidence in the payback. They can prepare a sophisticated analysis that very precisely identifies the investment and savings from each energy-efficiency technology. Real-estate firms could understand and master the entire field of energy efficiency, but hiring a department of energy experts is not the key to real-estate success. Successful real-estate firms must be good at designing pleasant and functional buildings. They need to be good at negotiating rents and helping their customers adapt spaces to their needs. They must excel at obtaining affordable financing for their buildings, since their profit is what they keep after paying the mortgages. Energy efficiency is very far down the list of skills with the potential to improve a landlord's profitability. This is especially true where laws forbid submetering and energy costs are passed through to tenants.

Energy efficiency is not a core competency for a typical property owner. Universities own much property, to help them carry out their mission of learning—of research and teaching. Universities don't focus on energy use. Did a student ever select a school by the quality of its boiler room? The energy efficiency of the university power plant is simply not a factor in college selection. It should come as no surprise that universities focus on their main missions of education and research. They simply aren't good at energy use. What is worse, they do not realize how inefficient their energy conversion and use are relative to optimal levels. They benchmark themselves against other universities who are equally inept at saving energy, and gain a false sense of competence. Senior executives of universities are selected based on academic credentials and administrative experience, so the university energy manager has little hope of ever becoming a senior executive in the university. Large hospitals and medical complexes have a similar situation. Management scholars have predicted that every function in a large institution which does not have a direct promotional route to senior management will soon be outsourced to firms which specialize in that function, and that this change will occur not just for savings but to improve the quality of service. We can expect major energy savings at universities and hospitals in the near future as they begin to hire specialist firms to manage their power plants and energy use.

Much of what follows is devoted to finding ways to change the total set of societal values and lead to choices that use less energy and save money. There may be a requirement someday to mandate fuel reduction that actually increases the cost of energy. But before reaching that step, we should pick the low-hanging fruit. Society should focus on understanding and fixing all the current laws, regulations, common practices, and attitudes that result in decisions to spend more money than is necessary and to use more energy than is required.

5

Linking Deregulation and Carbon Dioxide Reduction

THE U.K. EXPERIENCE PROVES THAT DEREGULATION SAVES CARBON DIOXIDE AND MONEY

On October 7, 1997, Mr. David Jefferies, chief executive officer of England's National Grid, speaking to an assemblage of electric utility CEOs in Boston, described the experience with deregulation in Great Britain since 1989 when the Thatcher government opened electric sales to competition. He stated, "Carbon dioxide per kilowatt hour has dropped 39 percent."[76]

Here was proof that we could reduce carbon dioxide without economic disruption, worldwide. The United Kingdom, under Margaret Thatcher, deregulated its entire electric market in 1989. The U.K. split the responsibility for distribution of electricity into twelve regional companies, opened up electric generation to everyone, and moved to retail choice by every electric consumer. By 1997, customers with as little as 100 kilowatts of peak demand could select their electric supplier on an hour-to-hour basis.

An interesting sidelight is the role that Margaret Thatcher played in overruling all the "experts" who said it could not be done or at least could not be done quickly. Sally Hunt, economist and vice president of National Economic Research Associates (NERA), was in charge of

NERA's consulting assignment to work out the details of deregulation for the United Kingdom. After considerable study into all of the intricate mechanisms needed to open the market, NERA consultants concluded that the timetables were unrealistic. Hunt reported to me that she personally told Prime Minister Thatcher of NERA's conclusion and Thatcher's reply was prompt and decisive. Thatcher said, in words to this effect, "We did not hire NERA to tell us if we could deregulate electricity. We hired you to tell us how. Please go back to work and keep the schedule."[77]

With this leadership, Britain did deregulate, on schedule, and has been enjoying environmental and economic benefits ever since. Reliability has not been compromised.

There was no public consideration given to reducing carbon dioxide by deregulating. The goal was to open markets and reduce the price of electric power to all sectors. The government simply said that the logic of monopoly was no longer valid and that its citizens would be better served by free markets.

The deregulation was experimental, did not fully open competition, and was greeted with deep skepticism for several years. As late as 1992, one observer wrote, "Under the current system, the prospects for energy efficiency are dismal," and the prospects for cleaner power plants "likewise dismal."[78]

Then market forces began to work. British carbon dioxide per kilowatt-hour of electricity dropped 39 percent in the first six years of free electric markets, without government mandates or incentives to use non-fossil technologies, without any mandated fuel switching, without any fines or tax credits. Once the state stopped using its police powers to prohibit competition, the magic of markets began. Self-interested competitors sought and found the generating plants that were cheapest to operate every hour, and these plants were, in general, more fuel efficient. The state retained ownership of the nuclear plants, but the handwriting was on the wall. Faced with competition, the operators of these state-owned nuclear plants experienced the motivation of fear and responded predictably. They improved their management and operating practices dramatically, resulting in higher availability. Power entrepreneurs have built 4,000 megawatts of small combined heat-and-power plants because these plants make cheaper electricity than electric-only plants. Power entrepreneurs have deployed 15,000 megawatts of combined cycle plants, are building another 9,100 megawatts, have automated existing plants, and in countless other ways have taken steps to reduce costs.[79] In the process, the use of carbon-based fuels has fallen dramatically.

Deregulation will work the same way in the United States. The U.S. electric market is much larger, and has an even broader range of technologies available. We are blessed with more hydropower opportunities and have vast areas with clear skies that are suitable for photovoltaic power. In general, we can expect U.S. deregulation to at least repeat the U.K. experience. With deregulation, newly freed competitors will simply find the cheapest producer of power each hour, with no thought for anything but cost (other than pollution regulations). Over time, power entrepreneurs will invest risk capital in new plants that fit their own view of the likely power future. Some will be right, some wrong, and some partially right. Those entrepreneurs who find lowest-cost solutions consistent with environmental rules will win. Others will follow rapidly, until the next even better idea is improved and perfected. This is the way of markets.

Governments now steer those choices in a variety of conflicting ways and use the power of law to stop competition. By reexamining laws, eliminating some and modernizing others to encourage efficiency as a pollution-control strategy, governments can lead society toward a sustainable energy future that improves the economy.

A RECIPE FOR NATIONAL ACTION

Ideally, the U.S. Congress will soon pass enlightened deregulation legislation that links lower costs with global climate-change-mitigation goals. All of the necessary parts have already been proven in other contexts and other countries. Here are suggested principles of deregulation that will insure a bright future with economic growth, fewer greenhouse-gas emissions, and less pollution.

End All Restrictions on Electric Generation, Distribution, and Sales

An ideal solution from the view of environment and consumer savings is to assure totally free movement of electricity to all Americans. The Constitution charges the federal government with ensuring the free movement of goods between the states to promote the general welfare. Electrons can certainly move everywhere with ease. Electricity is truly "natural" interstate commerce. No electric supplier should have any monopoly protection at all. Electric users should everywhere have the freedom to choose their electric supplier, whether they are presently served by investor-owned utilities, municipal owned utilities, Rural Electrification Associations, or federal power projects like the TVA (Tennessee Valley Authority) or Salt River. Free choice is available for groceries, houses, autos, clothes, travel, electric appliances, letter and package delivery, and almost every other commodity except the few remaining government-enforced monopolies. (Local telephone service remains a monopoly, but must be opened to competition soon.)

Opponents of federally mandated deregulation argue that electricity is a responsibility of each state, and a federal deregulation law would violate states' rights. These folks see national deregulation taking power away from the states. Once again, however, technology has overtaken law.

In the early days of electricity development, as noted in chapter 2, society wanted to encourage rapid electrification and chose monopoly protection as a way to induce power entrepreneurs to invest. Electric generation and distribution systems at that time were very localized and it made sense that individual states exercise the police power reserved to them under the Tenth Amendment of the Constitution to regulate the electric companies within their borders. Today, however, thanks to multi-state electric power grids, the production and distribution of electricity have expanded beyond the border of any one state into interstate commerce. All of the electric systems in the continental United States are joined in three grids including the western states, the eastern states, and Texas. Electricity generated in Ohio flows to New York and New England, while generation in Arizona powers Colorado factories.

Given the current interstate nature of the electric business, the commerce clause of the Constitution should now control and federal oversight should prevail. Dividing legislative supervisory authority between the federal government and the individual states in terms of electric transmission versus distribution or wholesale versus retail sale no longer makes sense. The Founding Fathers of the United States knew that they had to stop the individual states from protecting their "native" industries in order to encourage the fullest development of the national economy. Europe's long history of sovereign states practicing mercantilism and erecting trade barriers held back the full potential of Europe's economy until the European Union was formed and began to lessen trade barriers. If Europe can remove the trade barriers on electricity, surely it is time for the individual states in the United States to open their "electron" borders.

How to Double Electric-Generation Efficiency— Adopt a Fossil Fuel Efficiency Standard

It is possible and economical, using technology available in 1998, to double the fossil efficiency of the entire United States fleet of electric generation plants over the next twenty years and save money. This can be accomplished by gradually replacing old power plants that make electricity only with an entire range of plants that are either two to three times as efficient at converting fossil fuel to useful energy, or that use renewable energy.

The restructuring of the electric business should guide the thousands of present and future generators of electricity down a path of diminishing reliance on fossil fuel but let them choose among every possible technology. One way to accomplish this is to establish a national *Fossil Fuel Efficiency Standard*. (See chapter 9.)

The Department of Energy collects data from all utilities and tracks all fuel used to generate power and all power generated. It is easy to use these data to determine the fossil fuel used per average kilowatt-hour produced in the entire economy in 1997. Legislation can be enacted that establishes a standard of fossil fuel use per megawatt-hour of electricity that must be met, on average, by each producer. The legislation should then include a table for future years that lowers the allowable fossil fuel per kilowatt-hour over twenty years to one-half of today's levels. We need to give the economy time to shift intelligently to modern technology and to deploy any imaginable method to increase the electric production relative to fossil fuel use. It's important to make every generation firm meet the steadily tightening Fossil Fuel Efficiency Standard or pay penalties equal to the cost of the extra fuel burned. Let the market decide what works. We have calculated the fossil fuel use per megawatt-hour of electricity produced in the following way. For the national average, the numerator is all fossil fuel, measured in megawatt-hours, that was burned in a single year anywhere in the United States to generate electricity. The denominator is all of the megawatt-hours of electricity generated by any source—hydro, nuclear, biomass, fossil, wind—

Fig. 13. Historical U.S. Fossil Fuel Use per Megawatt
of Electricity Generated[81]

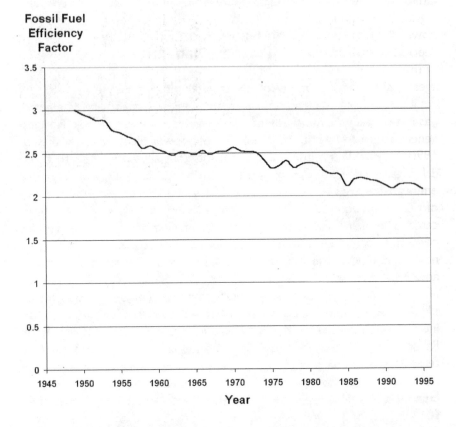

and adds the megawatt-hours of all heat recovered from electric generation and sold.[80]

Figure 13 tracks the megawatt-hours of the fossil fuel burned in the
United States to produce each megawatt-hour of electricity, and yields
several insights.

All of the regulated utility data is reported and collected, but private
power generation by combined heat-and-power plants was not reported
until 1989. The proportion of power produced with cogeneration was
roughly 25 percent of all power in the early 1900s, but fell steadily to
only 4 percent by 1978. If we had the data for 1949 through 1985, it

would not have much effect on the averages shown in the chart above, but would slightly lower the averages for this period. After the Public Utility Regulatory Policies Act (PURPA) was enacted in 1978, there was a gradual increase in the amount of cogeneration or combined heat and power built by independent power companies. IPPs are required to report fuel use and electric generation, but not heat recovery. Inclusion of this data, if it were available, would lower the fossil fuel efficiency averages more strongly in the last ten years than before 1985.

Even with this qualification, a very interesting picture emerges. There has been a steady decline in the use of fossil fuel per megawatt of utility-generated electricity. Over the forty-eight-year period from 1949 to 1996, the regulated utilities dropped their use of fossil fuel from 3.0 to 2.1 megawatt-hours/mWh of electricity. The big factor was nuclear energy, which did not exist in 1949, and which has risen to 20 percent of the nation's power.[82] A sustainable future will be easier to reach if we can develop acceptable ways to use more nuclear power. The current concerns with safety and disposal of nuclear waste may be addressed by new technologies which will render nuclear energy more cost effective and more socially acceptable. Until these improvements are developed and accepted, society will need to find other ways to use less fossil fuel.

A second factor in the falling fossil use per megawatt-hour was an increase in efficiency at fossil plants. There were some advances in Rankine-cycle plants and the new plants built after 1960 used higher steam temperatures and pressures which slightly increased average efficiencies.

Set Pollution Limits Per Unit of Output for All Electric Producers

All environmental standards today are based on input, not output, so they ignore efficiency. In order to unleash market forces in the energy sector, government must adopt consistent environmental standards for every generator of power, whether that generator is an existing plant or a new plant, and must base those standards on pollution per unit of energy produced. Neither is the case today. Currently, very tough standards are applied to new plants; older plants comply with the lower standards that were in force when the plant in question was built. Two-thirds of the power plants in service were built before 1972 when there were no environmental standards. Title IV of the 1990 amendments to the Clean Air

Act required some of the larger existing plants to reduce sulfur and NOx emissions to alleviate acid-rain precursors. However, it does not appear that this program significantly discouraged utilities from extending the life of old plants. Adopting output standards would reward efficiency and penalize inefficiency, unlike the present environmental standards that are based on fuel burned. This would open the door for the best environmental control strategy that there will ever be—the strategy of not burning the fuel. This change has become the main thrust of the Electric Power Supply Association. For detailed explanation, see Ellen Roy's persuasive article, "The Uniform Generation Performance Standard: Connecting Electric Industry Restructuring and Air Quality Improvement."[83]

Provide Transition Help to Monopolists

Transition payments should be provided to the utilities that have relied on the existing regulations to invest in generation, but only after allowing the market to determine what loss of value, if any, will be incurred in selling the total generating assets. This transition payment will assist the utilities toward modern and more efficient generation and eliminate the main argument against deregulation. Properly calculated, the transition payment makes the utilities whole, but does not create a windfall gain. The devil will be in the details, and several states have already crafted different approaches to determining and paying the fees for transition to competition. Unless there is a competition for the sale of monopoly-induced utility-generating plants that determines the best way to make those plants more competitive, utilities will overclaim their losses and consumers will pay too much. Given a chance, the market will decide the value of these plants. Existing generation-plant owners should be required to sell to the highest bidder all the plants they built under regulation. They could then claim transition fees to cover any net loss.

Massachusetts decided to force the electric utilities to sell all of their generating plants if they wanted to stay in the regulated electric transmission and retail distribution business. At this writing, the regulated utilities are in the process of selling their plants. New England Electric System (NEES) put all of its 5,000 megawatts of generation into an auction, and to everyone's surprise, received a winning bid of 145 percent of book value for the plants. This means that the winning bidder, U.S.

Generating, felt that the plants were capable of earning a stream of future cash that supported an investment of nearly one and one half times the book value of the plants. U.S. Generating saw value in excess of what was carried in NEES's books. Boston Edison has a similar experience, selling all of its generation to an independent power company, Sithe Energy, for 119 percent of book value. These Massachusetts utilities also had some contracts with non-utility generators to purchase power at prices that were above today's market prices per kilowatt-hour. These contracts were also sold, or more accurately stated, the utilities determined through auction the lowest price they would have to pay to some other party to take over the obligations to purchase the power. Other states have required the divestiture of some of the generating plants in order to create a local market for power, allowing the utility to keep some of its generating plants, but operating these plants without guarantee of price or rate of return.

I believe that the transition to competition can be made in a variety of ways, and we can already observe and evaluate many approaches. One possibility is a transition approach in which no utility would be forced to sell its generation plants, unless they claim transition fees; then they would be forced to sell every generating plant they own in an open market, to determine its true value. A third party could conduct these auctions and give the selling utility a chance to bid on one or more of the plants it is offering for sale. The idea is to let the market vote, with hard dollars, about the future value of every plant, then compare the total sale price for all plants with the book value of all plants to determine the transition fee. Some plants might be sold at a premium to book value, offsetting the sale of other plants and contracts at below book value. Consumers would be left with the lowest possible added charges to move to free markets, but the shareholders of every monopoly utility would not suffer loss due to the law change. This seems fair to all and avoids the intensive gaming that will go on if the transition costs are determined by regulatory hearings, as is happening in Pennsylvania.

Summary of Above Principles

This set of principles gives every firm a chance to compete and provides each a chance to find its own approach to generating power that is competitive and less damaging to the global environment. This set of principles provides for the maximum possible savings, because it lets the market find the cheapest generation at all times that is consistent with the gradual move to less fossil fuel use. This set of principles recognizes a societal responsibility for damage done when laws are changed, and suggests an approach to protect consumers' pocketbooks.

This approach of opening the markets to competition and subjecting every generation company to a Fossil Fuel Efficiency Standard mandatory schedule which steadily reduces the average amount of fossil fuel burned per megawatt-hour of electricity and heat produced sends some powerful signals to power entrepreneurs. It stresses their use of existing capacity as long as possible by either improving its efficiency or offsetting its fossil inefficiency with other parts of their generation mix. But each power company also knows that over time, they must halve their dependency on fossil fuel per megawatt-hour. This can be done by deploying combined heat-and-power and/or by converting to nonfossil generation that uses renewable or nuclear energy. There is time to improve the technology of burning biomass, time to lower photovoltaic costs, time to develop cheaper windmills, time to invest in equipment to recover and sell heat, and time to build new and more efficient generating plants. Advancing the technologies to reduce dependence on fossil energy sources will translate into corporate profits. This approach lets the market decide whether to increase the efficiency of fossil fuel plants, or to bet on renewable technologies as the best future solution. You can be sure that different entrepreneurs will pursue each possible strategy. There is a trend already well underway that lowers the cost of electricity made with renewable energy, and major energy companies are already making bets on the future of renewables. In part, they are betting that there will be government intervention to speed renewable deployment, like the Clinton administration's plan to require 5.5 percent of the electricity to be generated from renewables. But these companies are also seeing the potential to compete with the existing inefficient central electric-only generation. The *Economist* of April 18, 1998, made this observation:

> Between 1985 and 1995, installed costs of photovoltaics (PV) which convert sunlight into electricity, dropped by more than half, thanks to better design and greater efficiency. PV is now cheap enough to be competitive in areas that are not connected to the electricity grid. Wind power has done better still. Costs have roughly halved since

1990.... The result is that wind power is now just about competitive with fossil fuel in some markets.[84]

The government should not rule on which technology to use, but should adhere to two simple principles: (1) All generation of electricity and combined heat and power will be subject to the same pollution allowances per unit of output, regardless of when the power plant was built and regardless of the terms of its original air pollution permits. (2) All electric production and all heat recovery and sale from electric production, and all fossil fuel burned in any electric generation plant will be counted in calculating each firm or government utility's average fossil use per kilowatt-hour. All generators must reduce the megawatt-hours of fossil fuel they use per megawatt-hour of useful energy, or they must purchase credits from some other generator who beats the standard. (See chapter 9 for further discussion and comparison with other potential government approaches to reducing carbon dioxide emissions.)

Government research should move away from technology deployment—free market forces will rush to deploy everything that promises to meet tightening fossil fuel standards. Government research can instead focus on underlying physical principles upon which energy-conversion technology is based, confining its role to future developments that require patient capital. The emerging power generation giants, who will see worldwide opportunities to increase their profits by developing breakthrough technology, will also support basic research.

Lee Lynd, professor in the Thayer School of Engineering at Dartmouth College, summarizes this last point quite clearly. In personal conversations, Professor Lynd has told me that he feels industry is really good at near-term technology improvements, but is less likely to initiate long-term technology development. In his view, research labs are really good at dreaming up and improving long-term paradigm shifts, but really bad at near-term incremental improvements. Both should focus on their core competence.[85]

UTILITY STRANDED-ASSET CLAIMS PROVE ECONOMETRIC MODELS ARE WRONG

Most people seem to think that our fuel conversion has reached its peak. The assumption built into most economic models is that the United States is already using energy as efficiently as prices and technology allow. Given this assumption, many people do not expect any economic or fuel-saving gains from the deregulation advocated above. A raging debate is underway in most states and in Congress about how to deregulate electricity. While there are many competing views and approaches, the utility industry has been absolutely in agreement about one point—stranded assets. This interesting term is now applied to utility-owned generating plants which are so inefficient that they cannot compete. The cost of producing electricity in these plants is relatively high, but under monopoly regulation these costs are borne by all customers. Under free competition, new and more efficient generating plants will be built and will drive down the cost of producing electricity; the old plants will be forced to reduce prices or stop producing electricity. If the old plants reduce the prices they charge, they will not be able to pay down the related loans and equity. Hence, competition will strand these generating plants, or turn them into "stranded assets." Close examination of the claims of potential stranded assets proves that present fuel conversion is not optimal.

Every utility CEO has argued forcefully that his or her utility has plants that are uneconomic—that cannot compete against new, more efficient plants. These CEOs all say that their shareholders will suffer serious losses if monopoly protection is suddenly removed. They argue that competition will force them to accept lower prices for electricity from these inefficient plants, and the lower prices will not cover their costs or pay off their loans. Many CEOs claim competition is good and will work, but that there should be a recognition of the losses they will incur due to the law change, and they should be paid for these stranded assets as a precursor to competition in electricity.

I believe these utility CEOs are fundamentally correct. It is in their interest to overstate the argument and to inflate the claims of stranded assets in order to increase the transition payments they will receive, but there is a basis for significant stranded-asset claims. Many independent power companies have built and are building new plants that use far less fuel per megawatt-hour generated and are consequently cheaper to operate than many of the old plants now in use. Under competition many old generating plants will be forced by the market to close or to dramatically improve efficiency. There is little dispute about this from any part of the power industry.

Juxtapose this point with the widespread assumption that our fuel conversion is already optimal, given today's prices and technology. These are mutually exclusive possibilities, and energy professionals who are at risk for the consequences of opening markets are all saying that much of the present generation is not optimal. All of the reported economic models ignore the clear message of these professionals and assume that saving any carbon will cost money. The economic models are wrong. Their conclusions should not form the basis for policy until the models factor in the impact of competition. The entire electric industry agrees that it is not operating optimally. The industry knows it is generating power in many old, fuel-guzzling plants. Electricity from those plants is too expensive and cannot compete in the open market. These plants generate and produce needless CO_2 today because competition is blocked by monopoly regulations.

The damage done by these econometric models is immense, and one can only hope that a world-class economist will modify his or her model to include competition. Policy makers should reject any model that ignores facts which have been so clearly stated by the U.S. power industry and demonstrated by deregulation experience in the United Kingdom and Argentina. These corrected models would make the case for efficiency, showing how much competition could be expected to improve the economy and the environment. Policy makers need help in seeing how a 30 percent drop in electric prices over the next ten years might affect the economy. What is the economic value of reducing various pollutants by 50 to 95 percent, and simultaneously lowering electric prices?

A CLOSER LOOK AT POLLUTION REDUCTION

The technological development of power plants has focused on reducing regulated pollutants since the Clean Air Act was passed in 1972. In the ensuing twenty-six years, the progress in technology to reduce pollution has been awesome. If we build a new power plant to replace an old pre-Clean Air Act plant, levels of nitrous oxide can drop by 99 percent.

Figure 14 charts progress in reducing the NOx emissions of gas turbines since 1975, at which time a gas turbine emitted 200 parts per million of nitrous oxides (NOx). Between 1980 and 1987, dry low NOx and water injection, plus other advances in basic turbine design, dropped NOx to 25 parts per million. Since then, newer approaches have pushed the NOx emissions of commercial turbines down to 9 parts per million,

Fig. 14. History of NOx Controls for Gas Turbines[88]

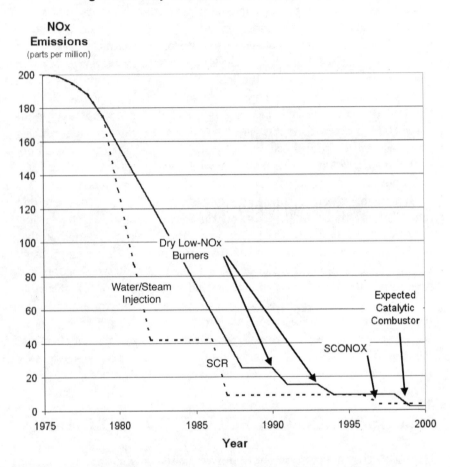

and California recently declared a new technology called SCONOx to be LAER, or Lowest Achievable Emission Rate. This process cuts NOx emissions to 3.5 parts per million by using catalysts which trap nitrogen, and release it in a regeneration cycle as elemental nitrogen—what it was in the inlet air.[86] A NASDAQ-listed company, Catalytica, expects to reduce NOx emissions to as low as 2 parts per million with its XONON® combustor which causes gas to ignite at temperatures below which any NOx is formed. The only NOx in the exhaust is formed in the small preburner. General Electric, Solar and Pratt & Whitney Canada are all working to incorporate XONON® combustion systems into their turbines.[87]

6

Facing the Challenge: Stabilizing Global Carbon Dioxide

☀ STATING THE REAL PROBLEM

Undersecretary of State Timothy Wirth led the United States effort on global-warming negotiations for several years. His access to top scientific resources was far-reaching and informed. At the National Marketplace for the Environment, a conference held in Washington, D.C., on November 18, 1997, Wirth laid out the challenge, paraphrased as follows.

Everyone seems focused on emissions of carbon dioxide relative to 1990. The media have reported the debate to be over whether we target a reduction of emissions to the 1990 levels, or to 5 percent below 1990 levels, or to 10 percent below. Wirth wryly noted that a person's virtue seems to be determined by where they are on this scale, with the most virtuous calling for the 10 percent reduction.

The issue, according to Undersecretary Wirth, was not emission levels of carbon dioxide, but the carbon dioxide concentration in the atmosphere. The atmospheric concentration has risen from 280 parts per million (ppm) in pre-industrial times to 360 ppm of carbon dioxide in 1997, and the scientific concern is to stabilize atmospheric levels somewhere between the present 360 ppm and double this level. It is, after all, the amount of carbon dioxide in the atmosphere that increases or decreases the reflection of infrared radiation back to Earth—that increases or decreases the atmosphere's greenhouse effect.

To stabilize carbon dioxide in the atmosphere at double the present levels, we must reduce our carbon emissions by 60 to 70 percent. The math is unfortunately very clear, and demonstrates the inexorable magnitude of our problem. Society will not lower emissions by 60 to 70 percent without major changes in its fundamental approach to energy use and conversion.

Tim Wirth's metaphor to explain the problem is a swimming pool (the atmosphere), which is partly full of water (carbon dioxide), and a pipe feeding water into the pool (carbon dioxide emissions). The pipe is flowing faster every year as society expands and consumes more fossil energy. These factors combined with a dependence on fossil fuel lead to more release of carbon to the atmosphere every year. If the flow of water in the pipe feeding the swimming pool increases, the swimming pool fills with water more rapidly.

To complete the metaphor, imagine that there is an open drain that allows some of the water to escape. Photosynthesis of plants removes carbon dioxide from Earth's atmosphere more or less as fast as the atmosphere gains carbon dioxide from plant decomposition. When Earth was without human influence, the level of atmospheric CO_2 changed very slowly. The maximum rate of change in CO_2 concentration over the last 160,000 years, as measured by the Vostock ice core samples, was just two parts per million per century.[89] The carbon dioxide released every year from plant decay, forest fires, volcanic eruptions, and other natural sources more or less adds carbon dioxide at about the same rate as the natural removal,[90] or sequestering of carbon dioxide by photosynthesis. Without human intervention, the carbon dioxide sequestering, or drainpipe, is keeping the carbon dioxide level nearly constant, or at least causing changes in carbon dioxide level to occur slowly.

Plant photosynthesis is almost solely responsible for the sequestering of carbon dioxide. Plants use water and energy from the sun to split atoms of CO_2 into carbon, which they use to build tissue, and oxygen. In the process, plants release oxygen. Animals eat carbon-containing plants, breathe in oxygen, and live off of the energy produced when carbon and oxygen join to form carbon dioxide. (This is obviously a great simplification, but it captures the essential relationships.) The carbon stays locked up as hydrocarbons, essentially sugars and cellulose, until the plant decays or is converted by animals and insects into methane (CH_4) or carbon dioxide. This sequesters the carbon removed from the atmosphere during the life of the plant and until the plant material decays. In other cases animals eat plant carbon and lock up some carbon as calcium carbonate, the stuff of seashells. Limestone deposits are actually storehouses of sequestered carbon, originally separated out of the atmosphere by photosynthesis. However, not all plant matter decays completely. Over eons, much plant material has been preserved as peat, coal, oil, and natural gas, and indeed as limestone. The amount of carbon that was removed from the atmosphere and sequestered has grown for at least 400 million years. The earliest fossil records of plants have been dated about 400 million years ago, and it can be assumed that from that point on, some of the plant remains ended up sequestered in peat bogs as precursors to oil, gas, and coal deposits. This geologic carbon storing started to reverse when humans learned to extract stored fossil fuel by building fires, liberating the sequestered carbon as carbon dioxide.

Returning to Tim Wirth's analysis, the problem is thus not carbon dioxide emissions, but stabilizing the level of carbon dioxide concentration in the atmosphere at somewhere between the present 360 ppm and 720 ppm. The problem is that as the level of greenhouse gas in the atmosphere rises, more heat is trapped and the average global temperatures rise, creating all kinds of havoc. Just one of the worrisome aspects of increased carbon dioxide concentration is that the slight increase in global temperatures caused by rising carbon dioxide concentrations could produce increased evaporation and more cloud cover, which would accelerate further warming and lead to violent weather changes. There is evidence from ice-core samples that the increase in temperature could be ten times as much as the climate models predict due to the feedback or associated effects of increased water vapor and less reflectivity or albedo because of decreasing ice cover.

I want to emphasize that no one is completely certain of the effects and the feedback loops that are a part of the global climate system, and there are certainly skeptics who argue that all is well. One group even argues that more CO_2 is a good thing, because it will make plants grow faster and open northern areas to crop production. Perhaps there will be some positive effects mixed in with negative effects of global warming. It is, however, very sobering to look at the weather of the first months of 1998, which were the warmest months on record.

It does not take much temperature increase to melt polar ice caps, and the current trend is exactly that—melting and breaking off of the ice shelf. On April 18, 1998, Ted Scambos, a research associate at the Cooperative Institute for Research in Environmental Sciences, a joint institute of the University of Colorado at Boulder and the National Oceanic and Atmospheric Administration (NOAA), posted this report on the Internet:

> Recent satellite images collected by the University of Colorado at Boulder-based National Snow and Ice Data Center indicate a section of a large ice shelf on the Antarctic Peninsula has broken away. The section of ice shelf that broke off was about 40 kilometers long and five kilometers wide. The satellite pictures appear to confirm earlier studies by the British Antarctic Survey that predicted the 12,000 square kilometer ice shelf was nearing its stability limit. Antarctic Peninsula ice shelves have been in rapid retreat for the last few decades, apparently in response to a regional climate warming of 2.5 degrees Centigrade, or 4.5 degrees Fahrenheit, since the 1940s. Ice shelves appear to be good bellwethers for climate change, since they respond to change within decades, rather than the years or centuries sometimes typical of other climate systems.[91]

By July of 1998, evidence of fundamental climate change became overwhelming. The *New York Times Sunday Magazine* ran an article on August 2, 1998 highlighting twenty-one separate highly unusual weather and climate events that occurred in the first seven months of 1998 and stated: "The year 1998 may be remembered as the time when weird weather became the norm. From the storms that pounded California over the winter to the drought that fed Florida's fires this summer, the one constant of the weather has been its bizarreness."[92] The article goes on to describe that once in a thousand years 21-inch rainfall fell in Santa Barbara, California, in February; the nearly one month of consecutive days of over 100 degree Fahrenheit weather in Texas; and the fact that eight of the ten warmest years in recorded history have been in the last decade. Daniel F. Becker of the Sierra Club stated, "The scientists are telling us that global warming means more extreme weather. The weather is telling us that global warming is here."[93]

With our increasing emissions of carbon dioxide, these are the dice we are rolling.

TECHNOLOGY SHOULD NOT BE THE MAIN FOCUS

Undersecretary Wirth's description of the problem is persuasive and based on access to top scientists. However, I do not think he has found the right prescription to tackle the problem. He has not offered insights that will let us mitigate climate change and not disrupt the economy. Wirth seems to believe the main answer is technology, but the technology needed to reduce carbon dioxide emissions by 30 to 40 percent already exists and with leadership to guide the power industry away from fossil fuels, we can do better. Today, all of these money-saving changes are held back by outmoded laws and governments preventing competition. Mitigating climate change without unleashing market forces is frankly beyond my imagination. On the surface, reducing fossil fuel usage by 60 to 70 percent seems impossible to achieve with even the most optimistic forecast of technology that is likely to be developed. A completely nuclear-powered economy would be one answer, but it does not seem to be politically and economically likely any time soon. It simply costs too much to make nuclear power. In 1988 an article in *Energy Policy* stated, "But since nuclear power is the costliest way to displace fossil fuels, every dollar spent on it displaces less climatic risk than would have been avoided by spending that same dollar on the best buys first."[94]

I am convinced that technological development should not be the main goal of a climate-mitigation strategy, and that technology development, taken alone, will not provide the payoff that is necessary to reduce emissions far enough and fast enough. Yes, technology can help. It can be developed and deployed more rapidly, and efforts should be made to do so—but by whom? Businesses seeking market share and profit growth, or seeking to avoid loss, have a long track record of cost-efficient technological development. If governments pursue this development as the key to climate change mitigation, I fear they will ignore focusing on the problems caused by existing rules, regulations, and laws. Governments must focus on the fundamental changes that must be made in every country to improve energy conversion and to inculcate values of energy conservation. Without changes in numerous laws at state and federal levels, we will almost surely fail to stop the increase in atmospheric carbon dioxide. In fact, until we loosen or rescind some of the legal barriers to efficiency and demonstrate that we can reduce fuel use and save money, it will be hard to gain a political consensus for mandatory reduction of carbon dioxide emissions.

WE NEED DETAILS BEFORE ADOPTING THE KYOTO PROTOCOL TARGETS

The present negotiating approach is almost certain to fail. Until negotiations have clearly specified details on how the world will cooperate, the Senate is unlikely to pass the Kyoto Protocol mandate for a 7 percent reduction of U.S. carbon emissions below 1990 levels. The treaty, without details, has little hope of being ratified by many developing countries. Washington and its elected and appointed officials are being asked to commit the United States to caps on national carbon dioxide emissions. Without detailing a specific plan of how to reduce carbon dioxide that does not, at the same time, disrupt the economy, agreement on caps will be difficult. One might think we would first agree to balance the budget, then negotiate what expenditures to cut and what to tax. In fact, it took the United States Congress and several administrations many years of growing national debt to gain agreement on a goal of a balanced budget, even though nearly all politicians understood the need. Each individual legislator was worried that budget balancing would involve cutting funds for a favorite program, or would involve passing unpopular taxes. As a result, the agreement on the balanced budget was delayed for years, until a plan with details was negotiated.

Agreeing to adopt the Kyoto protocol to cap carbon dioxide emission is similarly problematic. Legislators will ask how these caps are to be achieved, and they will want to know if capping emissions will harm their constituents. Lowering carbon dioxide by government action is frightening until one realizes the terrible inefficiency of our electric generation. In fact, many senators seem to believe that we live in the best of all possible worlds—that today's energy industry operates optimally. The current operating assumption of nearly every member of Congress, and many of the industrial companies, unions, and interest groups, is that capping carbon dioxide emissions will disrupt the economy, or lead to more pollution, or in some other way hurt their special interest. Does capping carbon dioxide mean wrecking the economy? Do caps force us to create a cumbersome command and control structure of Big Brother interfering in everyone's daily lives? Is climate change really just a scare tactic of people who believe in more government control, a scare tactic designed to achieve other agendas, like turning the economy over to environmental activists?

The Senate is unlikely to agree on carbon dioxide emission caps until the enforcement mechanism is more clearly spelled out. Every party to the debate is concerned about their constituents' competitive position with mandatory carbon dioxide reduction. Some fear U.S. industry will lose its competitive edge globally because they wrongly assume that lowering carbon dioxide will increase energy costs. Many developing countries have refused to accept caps because they are fearful of curtailing their economic growth. Their expected population growth and desired economic growth will depend upon much more energy use, and without higher energy-conversion efficiency and/or more economic ways to tap renewable energy, this desired growth will require more fossil fuel burning and more carbon dioxide.

In the United States each of the one hundred senators will vote to ratify a treaty only if they see the treatment as less expensive than the disease it will cure. A majority of our 512 legislators must agree, in the face of many vested interests, to change all of the laws that establish barriers to efficiency. In U.S. federalism, with its magnificent separation of powers, there are a vast number of elected and appointed officials who must make and agree on changes that eliminate barriers to efficiency and thus make meeting an emission cap possible. This number includes state legislators, regulatory officials, judges, and state and federal cabinet secretaries. Society must make very fundamental changes in the way it approaches energy to achieve 60 to 70 percent carbon dioxide emission reductions, and there is no realistic way to improve fossil efficiency without changes at the state, county, local, and even the individual level.

A big concern of energy entrepreneurs is that if the Senate agrees to caps, the carbon dioxide emission rights will be allocated on a historical basis, creating a windfall gain for the less efficient generating plants, and making it harder than ever to introduce competition and build efficient new plants. Virtually all of the United States emission regulations established so far have ignored efficiency and have granted pollution allowances based on the amount of fuel burned historically instead of allocating allowances based on output of useful energy. To enable the market to work and double the fossil efficiency of power production, it is essential that all pollution-control approaches be revised to provide allowances per unit of useful power produced. Society creates the demand for electricity and thermal energy. The power industry then generates the amounts of electricity and thermal energy demanded. Today, an energy entrepreneur wishing to build a new, more efficient plant may be required to purchase sulfur allowances and nitrous oxide allowances from an existing power producer who was awarded those allowances based on historical pollution. This slows progress by making new plants more expensive and rewarding old plants. If carbon dioxide is allocated the same way, based on historical emissions, then progress toward efficiency will be further delayed.

Until regulations are modernized to award pollution allowances based on useful energy output, it will continue to be very expensive to rebuild our power plant infrastructure and legislators will fear the costs to society of complying with the Kyoto greenhouse gas reduction targets.

Likewise, the world's nations will hold carbon dioxide emission caps in abeyance until implementation plans are well developed. The mechanisms of changing all societies' energy conversion and use are a vital part of each elected official's decision to accept binding caps. In fact, the approach of capping carbon dioxide emissions raises many questions about allocation, measurement, administration, and fairness.

CAUSES OF CARBON DIOXIDE EMISSIONS

Four interrelated factors determine how much human activities increase atmospheric carbon dioxide concentrations. They include: (1) population growth, (2) economic growth, (3) energy conversion efficiency, and (4) dependence on fossil fuel for energy. Hold any three factors constant but change the fourth and carbon dioxide emissions change. Governments may agree to take steps to increase energy-conversion efficiency, and may also increase reliance on renewable energy. Population-growth reduction is a charged political subject, and few elected officials will vote for actions they feel will limit their country's economic growth.

World-population growth must be addressed. Students of population growth are optimistic because the record shows the rate of population growth slows dramatically as people experience an improved standard of living. This suggests that developing countries need economic growth to solve the toughest of the four factors in carbon dioxide emissions—population growth. If we can induce greater energy-conversion efficiency and less dependence on fossil fuels in a way that promotes economic growth, we can expect a slowing of population growth rates.

CHANGING THE RULES TO LOWER CARBON DIOXIDE BY 60 PERCENT— SKETCHES OF A NEW PARADIGM

The United States can reduce fossil-fuel-based carbon dioxide emissions by 60 percent over the next thirty years without disrupting the economy. In fact, the changes that are needed will help the economy by reducing waste. Such claims require considerable proof, and such proof exists. There are examples that prove the merit of virtually every action we need to achieve these reductions.

The next three chapters focus on the major changes needed to cut emissions of carbon dioxide and not disrupt the economy. The changes fall into two broad categories—those affecting consumption of energy and those affecting the fuel-conversion industry. These two areas are completely different and are subject to different motivations. First, government needs to change its approach to energy for its own facilities and its rules about how the rest of society uses energy. The main focus of these changes should be to educate the public about the value of energy conservation and induce behavior changes. The next chapter offers a set of recommendations to achieve the educational and behavioral changes needed to reduce consumer's energy use. The second major change is to cause power producers to use fossil fuel more efficiently and to increase their reliance on renewable energy sources. Chapter 8 explores how to eliminate the barriers to deploying energy-efficient generating plants by modernizing laws and regulations. Chapter 9 looks at national interventions that could be made to induce the fuel-conversion industry to decarbonize.

As people examine these proposed changes, I do not believe they will find them to be disruptive to the overall economy. Certainly they require new thinking. They require eliminating existing rules and eliminating the use of state police powers to prevent competition in electricity. They call for ending subsidies that distort energy decisions. They call for modest changes that will inculcate an energy-conservation ethic in the body politic. These changes will create savings. They allow us to discard some very intrusive command and control policies of present environmental regulation and replace them with standards and market mechanisms. The changes must send strong educational signals. Democracy absolutely depends on an educated citizenry. Changes in the way we consume energy, and movement toward a sustainable energy future will occur only to the extent the public understands the issues and demands change.

Most important, the changes proposed here provide a basis for debate about the costs and benefits of reducing dependence on fossil fuel and thus moving toward a sustainable economy. By scheduling the changes to occur over time and mandating a gradual lessening of our dependence on fossil fuel, these policies give every company now engaged in energy production, energy conversion, or energy use a reasonable amount of time to change their strategy. Every company would have time to redeploy assets and resources to support strategies that would be rewarding in a future in which government guides us to sustainable development. Any entrepreneur with a dream about how to further increase efficiency would be allowed to compete freely. The recommended policies would let the marketplace decide hour by hour what is cheapest and best.

These proposed changes are as applicable to other countries, including developing countries, as to the United States. *Any country without disadvantaging to itself vis-a-vis others could adopt the proposed Fossil Fuel Efficiency Standard.* The beauty of this approach is that it does not disadvantage the United States versus other countries that do not adopt the standards. On the contrary, opening U.S. energy markets to competition and providing incentives to our businesses to develop and deploy more efficient technologies will strengthen the competitiveness of our firms. Other nations will quickly see this and will be forced to follow a similar path to remain competitive.

As a matter of fact, most of Europe has long had policies to encourage energy efficiency and less fuel use. All over Europe, energy taxes raise consumer fuel prices to double what they are in the United States, and not surprisingly, Europeans use far less energy. Yet, they do not seem to be at an economic disadvantage to the rest of the world. Adopting the changes proposed in this book in Europe will further lower Europe's fossil use without more taxes.

In all events, we seem to have only these choices:

- The head-in-the-sand approach: Ignore the president and top policy officials and hope they are wrong.

- The fairy-godmother approach. Hope some magic technologies will be developed to fix the carbon dioxide problem without forcing us to eliminate the inefficiency in our approach to energy conversion and use.

- The grow-up-and-get-real approach. Face up to the fact that we can do a lot better. Start with this list, add other ideas and seek consensus about the best course of action.

My choice is easy.

7

Changing Consumer Attitudes Toward Energy Use

Decarbonizing can occur in two ways that do not lower our standard of living. First, consumers can lessen their energy waste. Second, the fuel-conversion industry can lower fossil fuel use by converting fossil fuels more efficiently and using more renewable energy. We will address actions to impact the fuel conversion industry in chapter 9. This chapter addresses how to educate consumers and create incentives for them to waste less energy.

Energy-saving washing machines, inexpensive energy-conserving occupancy sensors that turn off the lights and cut the heat in unoccupied buildings, better light bulbs, and many other energy-waste-cutting technologies are proven and economical. Most people agree that these things are desirable, but how does society arrive at an active, energy-conscious attitude? How does society influence people's choices? Surely elements enter into these choices other than pure economic analysis.

The government can lead the consumer to focus on the issue of energy efficiency. It can lead with a message that encourages changes of attitude. It has successfully done so to reduce other environmental threats. It can accomplish the same objectives with energy use. Leadership can help reduce greenhouse-gas emissions. Government is in fact the antidote to the "Tragedy of the Commons" problem, and has long played the role of setting broad rules that define the limits of acceptable behavior. Governments determine that driving vehicles beyond a given speed will unacceptably increase the chance of major accidents, and thus set a speed limit. Government can and should also decide that energy waste will unacceptably increase the chance of global catastrophe and thus reward stewardship.

Asking government to set limits on energy waste and at the same time deregulate electricity may seem like a contradiction. It is not. Ending the use of police power to prevent competition in energy is necessary because the lack of competition results in excessive fossil fuel burning and the resultant pollution including greenhouse gases and acid rain. We must modernize environmental rules that currently penalize efficiency or fail to recognize efficiency as a pollution-control strategy. The original objectives of environmental rules to clean the air must be respected and even strengthened, but those objectives can be accomplished with an approach that also rewards efficiency.

Keep in mind our earlier observations about how markets work. Competition will drive any market to find the lowest costs of production for all inputs that have prices, but will not automatically pick the best environmental approach. Producing any good or service has a cost to the environment, but without government action, the producer does not see a price for the environmental impact. For example, a power plant that emits large amounts of sulfur dioxide causes acid rain which has been shown to damage trees and lakes. There is a cost to society of emitting the sulfur dioxide, but absent government action, no price is charged to emit the sulfur. In this situation, competitors will build sulfur emitting plants if the cost of production from those plants is cheaper than the costs from other cleaner power plants. This leads to a situation where the individual actions of competitors result in overall damage. What is missing is a price signal for the externalities, or external impacts of producing power. Governments can and do cause markets to minimize the external damage of production in several ways. Regulations can forbid production of excessive pollutants, can charge a fee or tax on the pollution, or can exert detailed control over the pollution control technology used by each producer. The third approach, which is common today, removes the ability of the market to make optimal decisions about how to reduce pollution. Market decisions are replaced by cumbersome bureaucratic approvals of each plant's construction and approvals of all major changes to each plant. This is an expensive way to meet the objective of less pollution.

We do not question the need for government to intervene in markets where the "Tragedy of the Commons" is likely to exist—where the completely unfettered actions of self-interested individuals have the potential to damage society. Government must craft intelligent guidelines that channel these actions to lessen environmental damage. However, there is a danger to the economy with the third approach of detailed regulation of each production plant. When government errs, it is usually because the rules freeze detailed approaches and bar market creativity, or because the rules are left in place after the original rationale is gone—made obsolete by new technology. Society will always need restrictions on acting purely in self-interest. The challenge is how to form those restrictions with the least possible limitation to creativity and market solutions.

What follows is a partial list of suggested measures that government could use to change public attitudes toward energy. The measures aim at improving our relationship with the planet, and accept that the proper role of government is to ensure that we will not irreparably damage the earth we leave behind for our children. Some of the measures were included in the Clinton administration's 1998 budget proposal to Congress.

NEW BUILDING CODES

- We should review the building code standards and enact guidelines, many of which should also apply to existing buildings, that force deployment of efficiency-producing technology to turn off lights when spaces are not occupied. This has been done in California. Every owner of a building should be forced to install setback thermostats that reduce the heating or cooling during unoccupied periods. Less waste translates directly into less fossil fuel burned.

- Standards should be kept as general as possible, allowing the market to deploy various approaches and solutions, but demanding that controls be in place by a certain date. Standards should be adjusted administratively to incorporate new technology. Laws should expire, or sunset, in ten years to ensure that legislatures revisit the issues at timely intervals.

- We should mandate use of occupancy sensors to turn off lights in empty halls, bathrooms, conference rooms, and classrooms. These devices use digital technology and internal microprocessors, are widely available today, and cut electric use in the particular space by over 50 percent. Compelling every building owner to deploy this type of technology will reduce building costs, and will slow the demand growth for new electric power. If the requirement remains general, competing technologies and firms can vie for the market to turn off unneeded lights.

TAX CREDIT FOR ENERGY-EFFICIENT APPLIANCE PURCHASE

The government has used tax credits in the past to induce consumers to change their behavior. In response to the energy crisis of the late 1970s, tax credits were given for a variety of investments by homeowners in energy-conserving insulation or better windows, and in solar collectors to capture renewable energy. Tax credits can now be used to speed deployment of efficient appliances.

Direct the Department of Energy to create and continually update a list of energy-efficient major appliances that exceed the average efficiency of the present stock of appliances by at least 50 percent. Include clothes washers, clothes dryers, and refrigerators. Offer federal tax credits to individuals who install a new appliance from the DOE's approved list of products which use at least 50 percent less energy per unit of work done than the national average. The DOE would update the list of acceptable appliances every time any manufacturer introduces a new or revised model that meets the DOE's efficiency tests.

The DOE would also review the overall stock of appliances operating in the United States each year and reestimate the nation's average efficiency for each category. As the years go by and people install more and more of the very efficient models, it will become tougher for an appliance to make or stay on the list. This new estimate of average efficiency would be used to increase the efficiency needed by a new appliance to qualify for the tax credit. Manufacturers will fight to gain the bragging rights of having appliances which are superefficient, and will continually find better ways to wash the clothes or dishes, to cool the food, or to vacuum the floor. It will take standards that progressively increase required efficiency to guide us to a sustainable future.

- The best clothes washers and dryers on the market today use 80 percent less energy than the typical models being sold; they also do a better job of washing clothes. Today's premier refrigerators are able to cut energy consumption at least in half. Government must encourage consumers to adopt more efficient machines.

- This individual tax credit attacks the largest consumers of energy in the home and sends a powerful signal to every citizen. If there is one thing Americans focus on, it is how to pay less tax.

- This tax-credit mechanism sends a signal to every maker of appliances that efficiency will be rewarded. It will unleash a flood of research and development. I would expect that the first models to qualify for tax credits would carry substantial price premiums over today's energy hogs. The learning curve, however, is always the same. As the volume of high-efficiency sales picks up and competition comes into play, the price of efficient appliances will decline steadily and many of the inefficient machines will cease to be available. The tax credits on windmills and photovoltaic devices generated a market for these renewable technologies. Various companies competed and drove prices down. Today, these technologies are competing in some applications without further tax credit.

ENCOURAGE DISTRICT ENERGY SYSTEM CONSTRUCTION

The massive development of district energy systems that move normally wasted energy from power plants and industrial processes to homes and buildings should be encouraged. These district energy systems are essential if we are to fully utilize the heat that will always be a by-product of electric generation with any type of fuel. These systems also eliminate the energy waste of keeping backup boilers on low-fuel standby in every building.

All technologies that use fuel to produce electricity have surplus heat. It is theoretically impossible to convert all of the energy in fuel to electricity by a combustion process. Consequently, an important way to increase electric-generation efficiency is to recover the waste heat, typically as low-temperature water or low-pressure steam, and use it to avoid burning other fuel—use the waste to heat buildings and provide heat for thermal processes. Examples of thermal processes are hot water for washing, steam to sterilize medical instruments in hospitals, heat to power absorption refrigeration machines that produce chilled water for cooling, plus all of the many uses of heat in production facilities. Nearly one-third of all fuel in the United States is burned to produce thermal energy to heat and cool space, or to process food, produce chemicals, make paper and cardboard, and even to make beer. To achieve maximum efficiency from electric generation, we must capture the normally wasted heat from power generation and use that heat for all of these thermal processes, thereby saving the fuel that is presently burned. To achieve these efficiencies, our society needs the infrastructure to distribute the heat, and needs a stock of buildings that can use the heat.

- District energy systems typically consist of underground, preinsulated pipes, buried beneath the streets and connected to the basements of major buildings on one end, and to power plants on the other end. Typical users of this waste heat include apartment complexes, commercial office buildings, medical centers, jails and prisons, government office buildings, convention centers, stadiums, and industry.

- Europe, with its high taxes on fuel, has relied extensively on district heating to reduce fossil fuel consumption. Typical European systems gather waste heat from municipal refuse plants, electric power plants, or even oil refineries. District energy accounts for 38 percent of the heating market in Sweden.[95] Nearly every city and town in Denmark has installed a district heating system that delivers waste heat and avoids burning more fossil fuel for heat.

- To ensure that large buildings stay warm, it is standard practice for all buildings without district energy to operate a lead boiler that produces the heat needed, and simultaneously operate a backup boiler. The backup boiler consumes a significant amount of fuel to stay warm and ready to automatically take over heat production if the lead boiler fails. The operation of these backup boilers in every building all winter is a huge waste of fuel. They are forced to operate at the lowest possible output level, where they may only convert 25 to 40 percent of their fuel to useful heat, and they are typically very polluting at these low operating levels. District energy systems have only one or two system backup boilers kept warm in an entire city, yet provide the same reliability. This saves fuel and reduces pollution.

• A nationwide mandate should be enacted requiring all new construction of housing and office space to be designed to use low-temperature hot-water heating systems, or to employ systems adaptable to low-temperature hot water. The hot water to heat these buildings can be produced by oil or gas boilers in the building, but these buildings can later be converted to use waste heat from a central system. If the building has been designed with electric baseboard heating, or with individual heat pumps in each room, it is nearly impossible to retrofit the building to use centrally supplied waste heat.

The importance of making buildings "ready" for district energy was demonstrated to us in Trenton, New Jersey, in 1984. A Holiday Inn had been built twenty years earlier with individual air conditioners in the wall of each hotel room. Very expensive heat came from electric resistance coils. (Electric resistance heating uses twice as much fuel as heating with boilers, and operating costs can be double. First costs for installation are lowest with electric resistance heaters in every room; it is cheaper to run wires than pipes. Pay now, or pay later, and pollute.) When the state of New Jersey converted the inn to an office building, they wanted to connect it to our district heating system and use the waste heat from a new 12-megawatt electric generating plant. However, this required that a complete new system of piping be installed throughout the building, which required tearing up walls in every room. The piping to carry heat from a central boiler or district energy system is easy to install when the building is new, but very expensive to retrofit.

Without central piping for heat and cooling, the building remains inefficient. Setting this type of standard prepares the stock of buildings for connection to sustainable systems when they are developed. Other approaches to heating systems in buildings such as through-the-wall heat pumps, or electric resistance heaters, make waste-heat use more difficult. Such rules do not constrain the market very much. It is still possible to provide the heat from a large electric-driven heat pump, from gas- or oil-fired boilers, or indeed, from solar collectors, all of which are more efficient than individual room heaters. These practices of insisting on district-energy compliant buildings exist in Scandinavia and have been a vital ingredient in their world-leading energy efficiency.

Another reason for making buildings district energy compliant is the existence of large quantities of low-grade heat that could be harnessed to heat buildings. For example, refineries burn roughly 8 percent of the crude oil they start with to heat the rest of the oil in order to vaporize the various constituents.[96] Almost all of the heat that was produced from burning part of the crude oil is transferred to the vaporized products, which have undergone a phase change from liquid to gas.[97] A phase change of any liquid to gas takes considerable energy, and that energy is available when the gas cools back to liquid. This is why a pot of water placed on the stove will begin to boil after a few minutes, but then continue to absorb the heat from the burner for an hour or more until all of the water has undergone a phase change to steam.

At a refinery, individual products are separated out of the petroleum by raising the temperature of the liquid to the point where the particular chemical vaporizes, or turns to gas. Jet fuel boils off as gas and is separated, but then all of the heat in the gaseous jet fuel must be removed to cool the gas back down to a liquid state and to make it cool enough to ship. The next time you pass an oil refinery, look for a long, low structure that seems to have very large can-type structures on top of the building, and is emitting clouds of water vapor. This is the cooling-tower facility where all of the heat is removed and rejected into the atmosphere. This waste of heat happens at virtually every U.S. refinery—the waste is not necessary. In Gothenburg, Sweden, a Shell refinery recovers the heat as low-temperature hot water, and that water is piped to the Gothenburg district heating system, providing heat for 250,000 houses as well as downtown commercial office buildings. In order to save a lot of carbon dioxide and money, it is essential to have a community-wide district heating system. It is hard to imagine a sustainable future that tolerates this waste of heat from every refinery.

ADD AN ENERGY-EFFICIENCY CRITERION TO FEDERAL SUPPORT

A simple way to change energy behavior without costing federal taxpayers anything is to make all federal support payment to universities,
housing units, and other users of energy contingent on the grantee institution achieving an increasingly tightened standard of energy efficiency.
For example, research grants to universities would be determined normally, but the university itself would first have to qualify as meeting the
standards of energy use per square foot of space. This approach treats
energy waste as the environmental problem that it is and forces energy-
efficiency decisions to be reviewed by the top managers of each institution. At present, these institutions do not focus on energy decisions, and
they preserve their available capital for mission-related activities.

The 3,500-odd universities and colleges and the 5,800 medical centers in the United States have, in general, very wasteful energy systems,
but also have a vital characteristic that enables energy conservation. They
almost all have central boiler plants and a district system of steam or hot-
water pipes. Some also have central chiller plants and a district system
of cooling pipes. They are perfect candidates for cogeneration or combined heat-and-power plants, because the heat is already provided centrally. They can also be vital "anchor tenants" for new district heating and
cooling systems. In fact, a number of major institutions like the University of Pennsylvania, Tufts University Medical Center, Harvard University,
Creighton University in Omaha, Nassau Community College on Long
Island (New York), and the University of Maryland at Baltimore are connected to district energy systems. In the case of the University of Pennsylvania, a new 150-megawatt combined heat-and-power plant produces
electricity and the normally wasted heat supplies the campus as well as
downtown Philadelphia, operating at 73 percent efficiency versus the
United States average electric-generation efficiency of 33 percent.[98]

This proposal to mandate energy efficiency does not force institutions to raise capital and enter the energy business. Energy companies
and energy entrepreneurs who offer unique ways to reduce energy use
are eager to invest capital and knowledge, and will install improvements
such as combined heat-and-power plants, connection to district energy
systems, or a host of automation and efficiency products that will reduce

fuel and energy use. The beauty of this approach is that it forces institutions to improve energy efficiency by deploying technologies that pay for themselves and generate profits. At the end of the day, these institutions will pay less for their energy, after taking account of the capital charge and profit to third parties. Nor does this suggestion force outsourcing of energy. If these institutions can improve their energy efficiency with internal staff and their own capital, fine. However, the government should withhold support unless the receiving schools, medical centers, housing units, and others find ways to reduce dependence on fossil fuel.

The standards for the institutions could be set by the DOE as energy used per square foot of buildings, or per bed for hospitals, or for some combination of measures that is fair to the various institutions that receive government support, but that guides each to better use of energy. The General Services Administration already sets energy use goals for the many buildings it manages, and this does change the behavior of GSA managers toward energy.

We saw an example of these standards at work in the decisions of the United States Mint in Philadelphia. The mint is a steam customer of Trigen Energy Corporation's Philadelphia subsidiary, and literally uses steam to make money—to make coins. They have for years taken steam at 150 pounds per square inch of pressure, then reduced the pressure with a PRV (pressure-reducing valve) to 15 pounds per square inch for use in the mint. The mint held a competition for their business and we suggested installing a backpressure turbine generator that reduces the pressure by "extracting" 200 kilowatts of electricity very efficiently. This wonderfully "green" device converts to electricity 97 percent of the heat energy it removes, offsetting three units of fossil fuel in an electric-only plant with only one extra unit of fuel in our plant. However, this approach felt like an innovation to the manager of the mint, had the perceived potential of causing trouble with the local monopoly electric utility, and after capital recovery, provided only modest dollar savings. It was the right decision for energy, pollution, and financial reasons, but was an easy decision to avoid.

The deciding factor for the mint was apparently the General Service Administration's requirements that the managers of GSA buildings improve their energy efficiency—reduce the energy used relative to previous years. This backpressure turbine generator set was essentially a substitution of capital for energy, and while it only reduced the net cost of energy slightly, it did save money, and it significantly reduced the energy used, and allowed the mint manager to comply with an enlightened regulation. Similar enlightened regulations can cause the many institutions in America to substitute capital—their capital or entrepreneur's capital—for energy. They will save some money and the environment will win.

In our experience, it is usually possible to reduce these institutions' energy use and save money, but often times the old guard in their energy plants fear job loss and fight any outsourcing or intervention by outside providers of energy services. Energy standards will force attention to broader issues than employee job comfort. Complying with energy-efficiency standards will, in all probability, reduce these institutions' costs.

A significant feature of this proposal is the educational impact. The mission of colleges and universities is to educate, yet their own approach to energy production and use is not good.[99] The stick of withholding grants unless energy efficiency improves will cause these institutions to focus on knowledge and applications of knowledge that will mitigate global climate change.

This approach will cause many institutions that receive funding from the federal government to become leaders in moving to a less fossil-dependent economy. Housing units, military bases, federal office buildings, and indeed any of the office buildings of a state or city that receive block grants or other federal subsidies will be forced to improve energy efficiency or suffer funding loss. These efficiency standards, set to increase every year, will send a clear signal to move toward a sustainable world.

END ALL FOSSIL FUEL SUBSIDIES

All subsidies for fossil fuel exploration and extraction should cease. To-date, we have justified these subsidies for energy-security reasons, and there will always be pressure on members of Congress from oil- and gas-producing states to provide federal support for local oil and gas industries. Such support has been justified as a way to manage the United States balance of payments or to make the United States less dependent on oil imports from countries with unstable or unreliable governments. All these reasons make sense only if one accepts the old paradigm and assumes that energy is optimally converted and used. It would be more cost effective to improve U.S. energy security and balance of payments by reducing fossil fuel use, which would drop imports from OPEC countries. Fossil fuel subsidies do not make sense in a world without room in the atmosphere to absorb very rapid release of four hundred million years of stored carbon.

It has been estimated that the United States provides direct subsidies to energy production on the order of six billion dollars per year, with roughly four billion dollars to fossil fuel, largely for depletion allowances. Oil and gas producers are given a tax credit based on the depletion of the reserves in the reservoirs they have drilled. This lowers the cost of domestic production by reducing the tax on profits and was justified as a way to encourage a domestic oil and gas industry. It is simply a subsidy on U.S. fossil fuel production.

One might think this subsidy would make fuel cheaper in the United States, since it lowers costs, but prices for fuel are set in world markets, based on worldwide supply and demand. In fact, the subsidies do not make fuel much cheaper in the United States, but make it relatively cheaper to produce fossil fuel in the United States. They spur local production that would not be economical without the subsidies. The benefit of these subsidies works its way into the world price of oil. These subsidies make it cheaper to find and produce oil and gas in the United States, so there is more supply, which slightly lowers the world oil prices. In turn, this encourages consumers all over the world to use more fossil fuel.

Subsidizing energy is a worldwide phenomenon. Figure 15 shows the results of a study of net subsidies/taxes by country and shows a correlation between net taxes and lower energy use per unit of Gross Domestic Product (GDP). Countries that have net subsidies of energy, like Venezuela and Mexico, are in general using more energy per unit of GDP than countries with net energy taxes like Japan, Germany, and France. South Korea shows a net tax position and high use of energy per unit of gross national product because it heavily taxes energy used by consumers, but does not tax energy used by heavy industry in order to encourage steel making and other energy-intensive industries.

Fuel subsidies were often conceived as a way to help poor people by lowering the price they pay for energy. Unfortunately, a fossil fuel subsidy to transfer wealth has the unintended consequence of inducing wasteful use of energy. The wealth transfer and support of poor people can be accomplished with direct grants, leaving the recipients to choose

Fig. 15. Commercial Energy Efficiency and Energy Prices[100]

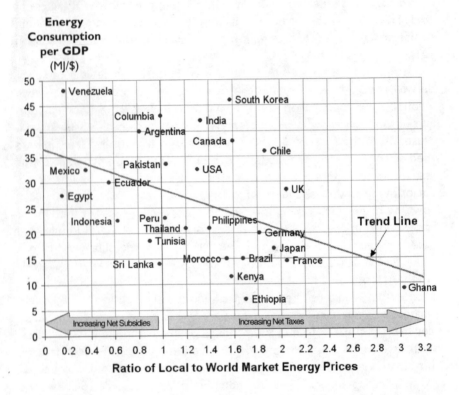

how to spend the money. The recipients would see the true cost of energy. All people, poor or rich, respond to pricing signals in making daily choices as simple as turning off a light and as complex as selecting an automobile, appliance, or new furnace. Ending fossil fuel subsidies will raise the cost of fossil fuel and cause consumers all over the world to see the full cost of energy in the purchase price. Passing through the true cost of oil and gas will have some effect in causing consumers to change their behavior and reduce fossil fuel use.

Enacting the full list of suggestions and many other ideas to guide people toward a sustainable energy future will end the energy-security problem. A staggeringly large reduction in demand for fossil fuel will result from these proposed changes to the United States energy paradigm. With the focus on a sustainable energy future, legitimate concerns about U.S. energy security will be met by lowered fossil dependence instead of expanded fossil fuel production.

This suggested end to fossil fuel subsidies should not be opposed out of false sympathy for the present oil and gas industry. These folks are energy professionals with a vital role to play in the sustainable energy future. They will make strategic adjustments based on a new set of rules and will deploy their considerable talents and resources toward earning a profit from the new opportunities. For example, the oil-refinery assets and skills form a splendid base on which to build a biomass-based ethanol industry. Whether automobiles run on fossil-based gasoline, or biomass-based ethanol, there is still a need to distill the product, ship it to markets, sell it to consumers, and so forth. All we need, in order not to disrupt our economy, is to give sufficient notice of the end of subsidies to allow fossil fuel extraction and refining corporations time to rethink strategy and redeploy assets. They will.

NATIONAL SHADE-TREE-PLANTING PROGRAM

A national shade-tree planting program should be created, with tax incentives and rules that apply to all property developers. Any tree sequesters carbon as it grows, but a properly placed shade trees shelters a house or a trailer from the summer sun, slows the heat-robbing winds in winter, and can reduce energy costs by 20 to 40 percent.[101] The fuel saved by reduced demand for heating and cooling of the house or trailer that is shaded by the tree is far more important to global warming than the carbon taken out of the atmosphere by the growing tree. For every one unit of carbon that the growing shade tree removes from the atmosphere to produce wood and leaves, ten units of carbon can be saved from the cooling and heating effect of the tree on the house.

A deciduous tree, such as an elm, oak, maple, or even the reintroduced chestnut species, automatically helps with the space conditioning of the building it shades. In summer, the leaves block the sun's heat and lower air-conditioning needs. In the fall the leaves drop, letting the winter sun through to help heat the house. Pioneers knew this and residents of older suburbs benefit from this phenomenon, but new housing developments and low-income developments typically cut down or do not plant trees. I shudder when passing a trailer park where there is not one tree;

the sun beats down on the aluminum trailers and turns them into solar ovens. This is very easy to remedy with trees and would add quality of life to the residents beyond the savings from less air conditioning. Composting the leaves creates a side benefit of healthy gardens the next year.

It is not terribly expensive to plant deciduous trees to the south of every house or trailer, or to require each developer to either save the existing trees, or plant new trees. This is an example of "Tragedy of the Commons." Housing developers often remove or do not plant trees because they want to avoid the small costs of raking leaves or of planting the trees, but the homeowners then pay for extra air conditioning and heating for the life of the properties, wasting energy. We recently signed a contract to operate the heating plant for Cherry Hill, a thirty-year-old project in Baltimore which houses low-income people and receives subsidies from the United States Housing and Urban Development administration. It is a reasonably pleasant campus of two-story apartment houses that serves 1,700 families, but there are hardly any trees. In the summer, residents are forced to turn on very inefficient window air conditioners or bake. It would be vastly cheaper to plant trees to the south of each building and pay to rake the leaves in the fall. And, it would make the housing complex much more pleasant for all the residents.

Planting trees will produce very economical energy conservation and improve the quality of life, but it has a larger purpose. It is a part of the educational effort needed to raise the entire population's focus on reducing fossil fuel dependence and on leaving a world habitable for our children.

TAX INCENTIVES FOR FUEL-EFFICIENT CARS

This suggestion mirrors the earlier notion of encouraging energy-efficient appliances with tax credits. We should authorize the Department of Transportation to identify the specific models of cars and trucks that have 50 percent greater fossil fuel efficiency than the fleet average for that class of vehicle and award the individuals who purchase these outstandingly efficient vehicles a tax credit of perhaps 10 percent of the purchase price. The calculation should be entirely based on fossil fuel, so electric cars would be charged for the fossil portion of electric production, currently 2 megawatt-hours of fossil fuel per megawatt-hour of electricity. By this calculation, an electric car is more fossil fuel efficient than a conventional internal-combustion-engine car. This suggestion makes much more sense in conjunction with a Fossil Fuel Efficiency Standard for electric power production, and would induce more widespread use of electric vehicles.

The tax credit should be used to signal consumers that fossil efficiency is important to the environment and to credit the environmental value of efficient cars. The standard could be recomputed every year to reflect changes in the fleet efficiency, so it is a moving target for the manufacturers to produce cars and trucks that are 50 percent more efficient than average.

There are many other major actions that could be taken to meet the challenge of reducing carbon dioxide emissions by 60 to 70 percent. However, the foregoing list contains important elements that need to be addressed to lower carbon emissions. Notice that there is not one word about research and development subsidies and little reliance on developing new technology. The proposed changes in rules and approaches create many potential rewards for people who develop and deploy new technology, and it is a virtual certainty that private energy-efficiency technology development will surge.

Many energy entrepreneurs will fight for the rewards and will drive the prices down on every product, passing the savings on to consumers. They will dream about great profits, but the reality of competition will force them to lower prices. The consumer and the future generations of our children will win. What is important is that proven, commercially available, cost-effective technology can greatly reduce our dependence on fossil fuel and our waste of energy. No new technology is needed to achieve significant reductions of energy use and greenhouse gases. New technology will augment what is already being stimulated by these proposals, however, and that will be a bonus.

8

Barriers to Efficiency

 ## "WHY AREN'T THESE MORE EFFICIENT ENERGY TECHNOLOGIES MORE WIDELY USED?"

Dr. Janet Yellen, Chair,
Council of Economic Advisors,
October 12, 1997

U.S. carbon dioxide emissions would drop significantly if the American economy simply deployed proven energy-efficiency innovations, especially those that convert fuel to heat and power. Throughout this book I have been describing some of the more prominent approaches to increasing efficiency and saving money. Why aren't the innovations installed more widely? Are the economic models right in their universal assumption that all economically justified investments in saving energy (and reducing carbon dioxide) have already been made? Are the only ways we can reduce fuel use and carbon dioxide emissions restricted to either taxing fuel to increase the price of inefficiency and/or stimulating new advanced technology?

Over the past twenty years, Trigen Energy Corporation and its predecessor companies have worked to find opportunities to earn a profit

146

by investing in energy conversion efficiency, principally by combining heat and power production. Independent power companies like our firm have found many, many opportunities. In fact, we seldom find a power plant, office building, hospital, hotel, university, or manufacturing facility where we cannot profitably reduce energy waste. Sadly, our attempts to improve efficiency have been repeatedly blocked both by the legally enforced monopoly of electric sales and by many other rules and regulations that are the result of a seventy- to ninety-year paradigm that forbids competition in electric supply. These barriers to efficiency extend into the very structure of environmental regulation; into tax policies, state laws, federal reimbursement and reward structures, and even labor laws. Most barriers to efficiency were erected by well-meaning people intent on achieving some social goal, but without regard to the damage the rules would do to energy efficiency.

The rationale for ever-larger electric generating plants that discard their heat has long ago become obsolete, but the accumulated burden of laws and regulations perpetuates that energy-inefficient approach. More troubling, these central plants were built with a self-interested bias by the utility to make them long-lasting and of a quality that was high for their time. This sounds right, but the technology of 1955 is not very good, no matter how well the plant was built that "locks in" that technology. They are so far away from other heat users that there is often little chance to even recover the waste heat and use it to save fuel elsewhere.

There is very little appreciation among theoretical economists, policy makers, or the general public of how many of these barriers exist, or how effective they are in blocking more efficient energy conversion and use. When fourteen companies were invited to discuss energy-conserving technologies with senior White House staff on October 12, 1997 (as discussed in chapter 4) and presented an incredible array of proven options, Dr. Janet Yellen, chairperson of the Council of Economic Advisors, asked the critical question, "Why aren't these more efficient energy technologies more widely used?"

This chapter responds to Dr. Yellen's question and to similar questions from Joe Romm, who until December 1997 was Acting Assistant Secretary for the Department of Energy's Office of Renewable Energy and Energy Efficiency.

Separate sections look at some of the barriers to efficiency in each of six different areas. The list is not comprehensive, but indicates how much the present approach to governance blocks entrepreneurs from building and operating more efficient plants and blocks manufacturers and utilities from profitably investing in end-use energy efficiency. We offer suggestions of how governance could be improved without sacrificing other societal objectives.

UNDERSTANDING BARRIERS TO EFFICIENCY

When any fuel is burned to generate electricity, no matter what the technology, there is significant heat left over. The best electric-only production technology today is a very sophisticated combined cycle plant that may convert 58 percent of the fuel energy to electricity. The combined cycle plant burns fossil fuel, usually natural gas, in a combustion turbine which drives a generator and converts as much as 42 percent of the fuel's energy content to electricity. Then the hot exhaust from the gas turbine is used to boil water in a heat-recovery boiler and produce steam, which steam is then used to drive a steam turbine generator. Up to 16 percent more of the energy in the fuel is thus converted to electricity, reaching a total of 58 percent of the fuel energy content converted to electricity. There is still 42 percent of the fuel's energy left over as heat. Scientists do not expect future fuel cells to convert more than 60 percent of the fuel to electricity, and the best present fuel cells are closer to 45 percent efficient.

Replacing today's inefficient electric generation with these new technologies would nearly double fuel efficiency and save money. But there would still be upward of 40 percent of the energy in the fossil fuel wasted as unused heat—heat dumped into the air or a river or lake. Combined heat-and-power generation captures most of the thermal energy and can achieve efficiencies of over 90 percent. However, much waste heat will still remain, which we can either use or lose. It is impossible to convert all of the energy in fuel to electricity because this would raise low-grade energy to high-grade energy and violate the laws of thermodynamics. The laws of thermodynamics apply to all machines that convert thermal energy into mechanical energy and heat. Whenever electricity is produced with mechanical energy, there will be heat left over. There are non-combustion methods of producing electricity including liberating fuel electrons in a fuel cell, but there is still no free lunch. A considerable portion of the energy content of the fuel will emerge as relatively low-grade energy, i.e., low-temperature heat.

As described in chapter 1, in the 1910s through 1930s local, state, and federal laws were enacted to encourage electrification of society and to promote economies of scale in electric generation. These laws achieved their goal of electrifying society, but they had an unintended consequence. The monopolies the laws created soon found they could earn acceptable profits and still throw away heat, the ever-present by-product from electric generation. Regulation allows them to waste two-thirds of the fuel and still be profitable. I cannot think of any competitive business that can waste two-thirds of its major raw material and even stay in business, much less earn a profit. Regulation simply fails to induce efficiency.

Before the enactment of these laws that made electric generation and distribution a monopoly, as much as 25 percent of the electricity produced in the United States was from plants that also recovered heat, or cogenerated. A competitive market would have steadily increased the percentage of plants producing and selling two products and fuel would have been saved. But society opted for efficiency-robbing monopoly protection, and by coincidence, real fuel prices steadily dropped over the next forty years, making it ever less important for electric generators to use all of the fuel energy. With the monopoly industry focus on electric-only approaches, technology was developed to raise the electric generation efficiency of ever-larger central plants from under 20 to 35 percent, but these plants still produced enormous quantities of waste heat. Nonetheless, in an environment of declining real fuel prices and increasing "Rankine-cycle" electric-generating-plant efficiencies, real electric prices fell.

With competition forbidden by law, there was little room (or incentive) for entrepreneurs with strategies to recover and sell heat to develop the "heat market" and gain market share. Consumers are generally not aware that there is massive energy waste in electricity production, which in turn results in excessive prices compared to what consumers would pay if the heat were also sold. Because of this ignorance, consumers have not objected to the inefficiency of old plants, or wasteful heat. Switching to doubly efficient central generation and combining the production of electricity and thermal energy whenever possible would be far more economical and would greatly reduce the burning of fossil fuel.

Technology Changes in the Mid-1960s

By the mid-1960s, large Rankine-cycle central-generation plant technology began to mature. Rankine cycle is the technology of making steam in a boiler and then using that steam to drive a steam turbine to turn an electric generator. Further gains in Rankine-cycle efficiency became elusive and excessively expensive. The final major step in increasing the efficiency of steam-turbine generation was a technological feat of heating the steam to "super-critical" temperatures, where it behaves differently. Developed in the late 1950s, the supercritical plants became increasingly sophisticated and temperamental. Nuclear plants using the Rankine cycle are less efficient (because radioactivity limits the pressure the pipes can safely contain), but lower fuel prices made such plants attractive. Both technologies are incredibly complex; such plants are frequently out of service for maintenance. The increasing capital cost of these large plants and their decreasing availability, began to drive real electric prices up in the late 1960s and throughout the 1970s.

Meanwhile, the technologies for smaller generation plants started improving. The efficiency of the best combustion turbines in 1970 was 22 to 25 percent, well below the 36 to 38 percent generating efficiency that large central boiler plants could achieve. Even after loss of 7 percent of the electricity in transmission, the large central plants still had efficiency advantages over these turbines, and the central plants could burn a solid fuel like coal, which was cheaper than the kerosene or jet fuel the combustion turbines required. But the gas turbine designs of the late 1960s have been improving steadily with no end to efficiency improvements in sight.

Today's Technologies Are Even Better

Because of technology changes and natural gas availability, the economic advantage today clearly goes to the combustion turbine. First, the efficiency of the turbine itself has now risen to 42 percent in larger machines of 150 to 250 megawatts. Second, modern gas-turbine plants combine two cycles to raise the electric-only efficiency to 55 percent or more. Current advertisements from turbine manufacturers have their sights on the gas turbine equivalent of the four-minute mile—60 percent efficiency. These plants first use the gas turbine to produce shaft power that turns one generator, then use the gas turbine's hot exhaust to boil water and make steam, which drives a second turbine-generator.

In addition, natural gas is now available almost everywhere at prices competitive with heavy residual oil. Gas prices are higher than coal, and will likely always be higher, since gas is an easier fuel to use. However, the nearly double efficiency from combined-cycle gas turbines and the lower capital and labor cost of gas turbines versus coal plants more than offsets the fuel-price difference. They are clean and doubly efficient. These gas-fired combined-cycle plants are the cheapest and quickest central plants to build today, and most new all-electric capacity being built or planned in the United States today involves such plants.[102] Society benefits every time one of these vastly improved plants is commissioned. However, under monopoly regulation we do not build these efficient plants to replace dirty old plants that are protected from competition; we build them only to meet new demands for power. The old plants are the problem.

The combined-cycle gas-turbine plants can be built in eighteen to twenty-four months for $450 per kilowatt versus seven years to construct a new coal plant at a cost of $900 to $1200 per kilowatt. Nuclear plants have taken eight to nineteen years to complete and range in cost from $1600/kW to over $5000/kW.

Another important consideration is that the large electric-only generating plants built from 1950 were frequently located far outside of cities. Their location makes it virtually impossible to recover thermal energy and use it for heating, cooling, and industrial processing. Today's efficient gas-turbine generators are smaller, cleaner, and quieter. They can be located within cities, close to customers. Excess thermal energy can be captured and piped to heat and cool buildings or to meet industrial-processing needs. Thus, the current technology can effectively provide two uses for the same amount of fuel. It can capture and use the thermal energy that is routinely wasted by the large electric-only plants. Our new 150 megawatt combined heat-and-power plant in Philadelphia is located in a 1907 electric plant and is supplying steam through thirty-three miles of buried pipe to provide thermal energy for roughly 250,000 people. Because of this location, we are achieving over 73 percent efficiency, twice the national average.

The perceptions of citizens about power plants are based on these old, large plants and hold back the deployment of center city combined heat-and-power plants. There is always "not in my back yard" (NIMBY) opposition to a new power plant because people picture a giant, ugly edifice with tall stacks. Yet no one objects to (or even knows) that every building has a basement boiler plant that burns fuel and emits pollution. Modern combined heat-and-power plants emit less pollution than the boilers they replace and cut fuel use in half. There is a need to educate people so they won't fight against the construction of these efficient combined heat-and-power plants near where they live and work. To recover and use waste heat, we must locate these small, clean power plants near housing and offices.

In 1980, when Trigen was developing a new district heating system to serve Trenton, New Jersey, civic groups opposed siting a power plant downtown, fearing it would be ugly and polluting. With considerable public relations effort and cost, we were finally allowed to build on a redevelopment site in center city. The new plant replaced many poorly run boilers with short stacks, and ground-level pollution decreased at the Environmental Protection Agency's Trenton monitoring stations. But to feel the real irony, take the Amtrak train to Trenton, as I do. Walk out of the station to the cabstand and say to the driver, "Take me to the downtown electric power plant." Invariably, the driver gives me a blank stare. Sometimes they ask if I mean Duck Island, which is four miles out of town and the site of a 1950s central electric plant that makes 600 megawatts. I say, "No, it is right across from the Labor and Industry building." Typically, the driver then says, "Oh, is that building a power plant? I wondered what it was."

A competitive market for electricity will rapidly deploy many of these combined cycle plants, and will drive the old inefficient power plants out of operation long before they die of old age. This is, contrary to what some claim, a good result, and will save consumers money. It is the pattern in every competitive industry and has led to continued economic growth and lower real costs of most products.

One dramatic proof of the value of these new electric generation technologies is found in the recent United Kingdom experience. In the five years after the 1989 deregulation of the electric market, combined cycle plants have captured increased market share and created dramatic efficiency improvements. The result has been a reported 15 percent savings for the average retail electric customer. Carbon dioxide from U.K. electric production has been reduced 39 percent since 1989.

Monopoly Rules Slow Adoption of New Technology: You Get What You Reward

Under electric monopoly rules, market forces are forbidden, and ponderous regulatory proceedings are required to gain permission to build a new plant. The regulators don't focus on increasing the efficiency of existing plants. Their job is to ensure that utility profits do not exceed a reasonable return on capital. They assume that the optimal course is to build new plants only if more capacity is needed, regardless of the efficiency or cost of operating the old facility. When capacity is needed, regulators pay some attention to what the best choices are, but to my knowledge they have never, in any state, ordered the utility to look for ways to recover and sell waste heat.

This approach by regulators, who focus on limiting profits, has produced numerous unexpected and unfortunate results. First and foremost, utilities keep generating capacity in service as long as it is capable of operating, even though they know how to build new capacity that is more efficient, less polluting, and cheaper to operate. Hampered by the regulatory miasma, we keep dirty, expensive plants operating as long as possible and new, more efficient plants are not built until the capacity is needed. If this logic were applied to computers, there would be very few personal computers in service today, because our society would still be maintaining the old mainframes. The mistaken regulatory logic would deem it unnecessary for Michael Dell or Compaq to build a new personal computer plant until the last IBM or UNIVAC mainframe factory was defunct. The same monopoly logic was at work for many years in the telephone business: the 1950s rotary phone in my boyhood home was still circling through the 1970s. When competition was allowed in making phones, markets responded. We now enjoy portable phones with touch-tone dialing, built-in answering machines, and a host of other features. And competition will transform power plants in the same way.

The monopoly regulators have also unwittingly reinforced an erroneous idea. As society begins to deregulate electricity, there remains a deeply ingrained attitude that central plants producing only electricity and wasting the by-product heat are the best economic approach. I don't believe that this is the case. The success of independent power companies (IPPs) in saving customers money on energy and earning a profit belies this large central generation idea, but most of today's environmental, siting, and tax rules are based on these mistaken central-plant-is-best ideas. Unless the rules are rewritten or rescinded, we won't realize the economic benefits possible from today's technology.

Excessive Rules Resulted in Ever-Larger Central Plants

Most of the successful independent power companies, looking at the sea of regulations they have to satisfy to build a new plant, have opted to focus on large, electric-only combined-cycle plants. PURPA opened the door to electric generation by independently owned plants. Such plants had to sell at least 10 percent of the energy in fuel as heat and be 1.5 times as efficient as the national average. In response, the IPP community simply built electric plants as large as the heat load would justify even though it would have been more fuel-efficient to build a smaller electric plant that could use all of its waste heat to serve the customer. Until the early 1990s, independent power companies, including the fifty or so subsidiaries of regulated electric utilities, could make more money building larger electric plants than by matching the thermal and electric load of users for optimal efficiency and thus minimize carbon dioxide production.

I recently spoke with Roger Sant, the chairman of AES. He founded AES in 1983 and it has become one of the most successful independent power companies and has gained a worldwide presence in fifteen years. AES's investment in power plants has recently reached $10 billion, making them one of the most phenomenal startup companies of all time. The executives at AES clearly know how to make money by producing electricity. Yet, Roger said, "Tom, you and I agree about the value of deregulation, but we disagree about the value of heat."

Roger is both knowledgeable and a good businessman. He has also always been socially responsible. AES spent $2 million in 1989 alone to plant over fifty million trees in Guatemala to reduce carbon dioxide.[103] AES improves the efficiency of every plant it buys or builds versus conventional monopoly production. Nonetheless, Roger and his team have looked at all of the present rules and regulations, and concluded that AES should remain focused on electric-only plants, which will still waste at least 40 percent of the fuel energy as heat. The decision calculus of Roger Sant and his colleagues and peers in independent power companies will not change until the regulatory apparatus is radically altered and until environmental rules reward efficiency.

Small-Scale Power Generation Improves

There has been one final significant technology trend. Cogeneration near heat users has become competitive with large central generation. The efficiency of smaller gas turbines has improved significantly and has become much more economic. New turbine technologies, which were developed for aircraft and large stationary power turbines, are being adapted to medium-sized turbines in the 500-kilowatt to 10-megawatt range, and even smaller sizes to fit the needs of small commercial operations. Several major companies including Allied Signal and Capstone have recently announced plans to produce and market gas turbines with as little as 25 to 75 kilowatts of electric output.

However, today's market for new electric capacity is dominated by demand for very large-scale central electric-generating plants. There is a chicken-and-egg problem that leaves large plants relatively more economical to build than smaller plants. The tax, environmental, and regulatory rules assume all electric plants are large, and penalize small plants. The result is a relatively low demand for extremely efficient smaller plants that are sized according to the heat needed. Low demand keeps capital costs high for small power plants. The two largest manufacturers of medium-sized gas-turbine generator sets (gensets), Solar and Kawasaki, each produce only about one gas turbine genset per day.

The manufacturing process for gas-turbine generator sets remains very labor-intensive and expensive. I recently visited Kawasaki's works near Kobe, Japan, and was struck by the contrast between Kawasaki's one-generator-set-per-day production in the gas-turbine works and their one-motorcycle-per-minute production in an adjacent factory. Clearly, mass production could greatly lower the cost of producing gas-turbine-generator sets. As rules change to encourage more energy-efficient cogeneration, the market demand for smaller engines and turbines will increase and prices per installed kilowatt will drop. Higher volume will also speed up technology development. It is difficult to mass-produce very large gas-turbine power plants; they will never have fully automated production. It is possible in the next ten years to see mass-produced combined heat-and-power plants with installed costs that are lower than all other alternatives.

Natural gas availability has made it possible to deploy reciprocating engines and combustion turbines everywhere, close to heat users or to heat networks such as university campuses or medical centers. Emission-control technology for gas turbines has made enormous progress, dropping regulated pollutants as much as 99 percent since 1975.[104] Rapid deployment of these small plants will increase efficiency and reduce cost of energy while improving urban air quality.

The technological advances are not limited to gas-fired turbines. Fluid-bed-coal technology and advances in wood chip boilers have enabled clean and efficient combustion of solid fuels, including coal, in small plants where heat can be recovered and used. There are over a thousand industrial sites that now burn some type of biomass to produce heat for their processes. Trigen owns and operates biomass-powered plants that serve Cargill (recently sold to A. C. Humko, burning rice hulls), Broyhill Furniture (scrap wood), Sara Lee (wood chips) and Kimberly Clark, Baxter and Gillman Paper (sawdust and wood chips). There are many more potential sites where biomass can replace fossil fuel for steam and some electric production. In Burlington, Vermont, the municipal utility operates a 50-megawatt electric-only plant that uses wood chips for fuel and could send its waste heat to replace the boilers at the University of Vermont campus.

Almost every firm or institution that owns multiple buildings or factories on a contiguous site has installed a central-heating plant and systems of pipes to distribute steam or hot water to the buildings. An example is Coors Brewery in Golden, Colorado, the largest single site brewery in the world (20 million barrels of beer per year). Coors's facilities stretch five miles from end to end, but are all served from a central power plant—in this case, an efficient combined heat-and-power plant. This system of centrally producing and distributing steam to all of the company's or institution's buildings is typical; using combined heat and power plants to produce electricity and steam is less typical today, held back by many barriers we have discussed. The DOE has estimated that there are over six thousand nonindustrial central-heating systems, and most are ideal for cogeneration.

Trigen recently participated in a DOE-led effort to identify the potential for new electric generation that would be sited next to existing steam systems serving institutions and industries. The conclusion, consistent with our internal studies, was that the United States has the potential to build over 120,000 megawatts of very efficient combined heat-and-power plants just to supply the waste heat to serve existing, already assembled steam loads. This would replace 13 percent of the 880,000 megawatts of total U.S. installed electric-generating capacity, save an enormous amount of fuel, reduce pollution, and reduce costs. These new plants using proven technology emit about 5 percent of the pollutants of the plants they would replace.

Inefficient Fuel Use Creates Problems

In the early 1900s over 25 percent of the nation's electricity was produced in cogeneration plants. By 1978 only 4 percent came from such plants. Energy efficiency simply was not encouraged by our monopoly system, but problems were beginning in the 1970s. One early problem caused by our bifurcated and wasteful energy-generation system was the heightened demand for oil, a demand that was partially responsible for the OPEC price shocks of 1973 and 1979. OPEC oil-price and availability shocks triggered a short-lived American focus on efficiency in the late 1970s and early 1980s, but then new oil and gas discoveries and dramatically improved drilling technology increased fossil fuel supply and lowered the real price of oil and gas. Look at the phenomenal popularity of sport-utility vehicles over the past few years and you know that energy profligacy has returned.

BARRIERS TO EFFICIENCY

Independent power companies and owners of heat-using facilities face an overwhelming array of barriers to deploying efficient cogeneration. These barriers fall into seven classes, including:

- laws or regulations to protect monopoly
- condoned anticompetitive practices of monopolists
- environmental laws that ignore efficiency or assume central generation is optimal
- regulations that prohibit optimizing plant operations with the grid
- tax policy

- federal reimbursement and reward structures

- state laws based on obsolete technology

The rest of this chapter looks at these barriers in detail and suggests ways to remove them and still meet society's goals.

Laws and Regulations to Protect Monopoly

There have been laws in every state since 1930 or earlier that make providing electricity to the public subject to regulation by the state public service commission. These laws establish state public utility commissions and empower them to approve rates, capital expenditures, financing, and other operations of the regulated utilities. The state laws typically give commissions power to award an exclusive franchise. There are exceptions. In Oklahoma, the voters must approve a franchise, and Ohio permits cities to have overlapping distribution franchises so that both an investor-owned utility and a municipal utility may serve the same customers. This is also the case in Lubbock, Texas.[105]

Some state laws also regulate steam distribution, not because it is a natural monopoly enterprise but because early steam systems were owned by the regulated electric utility. Since those systems obtained their steam from electric generating plants, the commissions felt they had to include the regulation of steam systems in order to make sure the electric utility did not allocate most of its costs to electricity, where rates of return were regulated. This would let the utility earn large profits by selling "free" steam from an unregulated subsidiary. So steam was regulated. The logic of regulation disappears when nonelectric utility companies run steam systems, but the old regulation seems to stay in force. (An independent power company charges what the market will bear for each of its products—electricity, heat, and chilled water. There is no issue with one class of customers paying a subsidy to other classes of customers and hence no logic for a commission to regulate the prices charged.) Heating buildings is hardly a monopoly, as one can employ gas- or oil-fired boilers, use electric heat, or purchase steam from others, but old laws die slowly. Many states do not regulate steam systems, because there was never an electric-utility-owned steam system in their state. In other states like Pennsylvania, the district energy systems are no longer owned by the regulated electric utilities, but the legislators have not rescinded the law. Regulation continues to the detriment of efficiency.

Regulated Monopolies Have No Incentives to Be Efficient

Each state regulatory commission's main task is to make sure the monopolists won't take advantage of their market control and make excessive profits. Many people on commission staffs, as well as the public advocate and consumer groups, who intervene in rate hearings, seem to blindly focus on the utilities' profits as the only way to lower consumer rates. The profits usually represent only 4 to 6 percent of the total price charged, whereas fuel represents 45 to 55 percent of the total price charged. This focus on preventing excess profits ignores the real money savings from increasing efficiency. This approach focuses on 4 percent of the cost (profits) and ignores 45 percent of the cost (fuel).

The typical commission procedure has the unintended consequence of curtailing innovation and efficiency. To prevent excess profit, the commissions have almost always mandated that the electric or steam customers will receive all savings from operations, but also have allowed the utility to "pass through" all fuel costs. This ubiquitous approach to regulation removes any incentive to increase efficiency. In most states, electric users automatically pay a fuel charge equal to the fuel burned to make one kilowatt-hour, and this is multiplied by the price the utility paid for that fuel. In most states this is done automatically by a "fuel adjustment clause." (Missouri requires a rate case before utilities can pass through the cost of fuel.)

The message to utilities is clear. If they invest in efficiency, or recover some of the heat they are presently wasting, they may earn praise, but no money. To make any profit from the efficiency investment, the utility must submit a rate case, show that they have made prudent investments in efficiency, and seek to increase their base rates to cover the return on the new capital. This hearing takes eighteen to twenty-four months and is required to include all aspects of the utility's business. There is no certainty of commission approval of a base-rate increase to cover the new capital, even though overall rates will have dropped automatically due to fuel savings. Under these rules, it is easier to ignore efficiency, and this is precisely what the utilities all over the United States have done.

The average efficiency of delivered electricity for the United States has remained at 33 percent for four decades. Compare that to other industries that have applied technology to reduce their energy use per unit of production. Compare this track record to personal computers that have doubled processing speeds every eighteen months with price decreases. This complete lack of progress in national electric-generation efficiency is the strongest possible proof that our regulatory system is broken.

I was presented with a vivid example of this mentality when Trigen purchased the Philadelphia steam system, whose rates are regulated. As I toured the power plant with Bob Reike, Philadelphia's vice president of operations, I noticed several areas of inefficiency. At each, I would stop and ask, "Why don't you change this process to reduce fuel?" Twice the answer was: "Tom, you don't understand. Fuel is a pass through and we cannot keep any of the savings." By the third time, I had made up my mind that Bob's attitude, even though a perfectly logical response to the rules in force, would jeopardize our company's long-term future because we were not a monopoly. The gas companies, oil dealers, and electric companies all try to lure our heating customers away with lower prices. We have to continuously improve efficiency and lower costs or lose business, because competition is vicious. Shielded from this lesson, the utilities can neither lose business by tolerating inefficiency nor gain a reward from improving it. The soundest rule of life may be that offered by Bob Nelson, a management consultant who said, "You get what you reward.[106]

This story has a not-so-nice epilogue. In 1998, we made a decision to stop approving further capital investments for efficiency in our Philadelphia system. All of the savings go to users and we simply cannot afford to make more "donations." In late 1997 Pennsylvania adopted a plan to deregulate both electric production and sale during the 1999–2001 period, but has so far left district steam subject to regulation, thus blocking efficiency investments. The process that we must go through to gain a rate increase is designed for a multibillion electric monopoly, and is very expensive for our one-hundred-person Philadelphia system.

Paradoxically, the flip side of this regulatory process that does not reward efficiency is that all utilities have overinvested in plant, which raises rates and thus profits. They have sought rewards under ill-conceived rules. First, they persuade their regulatory commissions that to achieve better reliability, they need to have larger reserve margins—i.e., more spare generating and transmission capacity. Next, they justify (even to themselves) that it is optimal to build every plant with very high-quality design so it will last a long time. High-quality construction costs a lot, and because these high costs go into the rate base and are allowed to earn a profit by the regulators, high costs are good for the utility. This perverse high-quality construction produces long plant lives, which seems like a good idea. The trouble is that longevity of big plants tends to freeze technology. They are hard to change as technology improves and seem too expensive to replace. Most of these plants have become old white elephants. They were built with enough quality to run as designed for fifty years, but the design only achieves 30 to 35 percent efficiency and is very polluting. New plants, based on today's technology, achieve 55 percent efficiency and cut emissions by 90 to 95 percent. Even though both carbon dioxide and cost could be cut in half by systematically replacing the old and inefficient generating plants, the regulatory commissions do not force the utilities to consider early retirement of plants.

So why have utilities found it easier to overinvest in new plant and transmission, but not in efficiency? The answer is scale. Investing small money in efficiency improvements requires a cumbersome process of hearings to gain approval to raise rates to cover the investment. Raising rates to cover very large investments in a new generating plant requires the same cumbersome process. It requires a set of regulatory hearings (which can take eighteen months) to gain approval to build. Why not gain approval for a huge, multimillion- or multibillion-dollar investment instead of for a few million dollars of efficiency improvements? As the new plant goes into service, the utility is allowed to raise its rates to cover all the debt costs and earn a 12 to 15 percent return on the huge new investment. This is a much easier way to make money than to fiddle endlessly with small efficiency improvements. And yet, endless fiddling is the hallmark of an efficient, market-driven economy.

ACTION: The best action to cure the problems caused by states enforcing a ban against competition in electricity is to mandate open access for all Americans to the electricity supplier of their choice, and let markets control profits. This would mean the total elimination of regulations on electric generation and sales. This will require the development of transition rules to help the utilities that made investments based on the old rules. Utility management teams could then focus on reducing costs through fuel savings to increase their profits. In the process utilities will begin to explore the combined production of heat and power, and will begin to replace old, polluting, inefficient plants with modern clean plants.

Retail Electric Sales by Competitors Prohibited

If any person or firm other than the monopoly utility considers investing in an efficient combined heat-and-power plant, the economics will be driven primarily by the expected sale price of the heat and electricity. In most, but not all states, it is legal for a third party to own that plant. A third party is someone other than the person or firm that owns the facilities to be served.

Some states like Louisiana, Maryland, and North Carolina go even further to protect their local monopolies. They have made it illegal for an entrepreneur to build a power plant unless all of the power and heat will be used strictly in his or her plant and not sold to others. Why? Most hospitals, universities, and industries are not in the business of converting fuel to electricity and they do not want to tie up their capital, or to hire and maintain the internal expertise needed to operate sophisticated combined heat-and-power plants. These states ban independent power from investing in or owning these efficient power plants. Thus, inefficiency continues.

In most of the states that do allow a third party to build and operate a power plant, the biggest economic issue is the price at which the plant can sell its electricity. Before 1998, the laws in all states denied retail sales and allowed non-utility plants to sell electricity only to their competitor—the local electric utility—at wholesale prices. These wholesale prices today are typically between two and three cents per kilowatt-hour, versus average retail prices of roughly six to nine cents per kilowatt-hour. The Energy Policy Act (EPACT) of 1992 opened wholesale power sales to everyone, but left retail sales of electricity under the purview of each state—still a protected monopoly. The 1992 EPACT has had a minimal impact on efficiency and has not resulted in replacing old inefficient generation, as explained below.

Every regulated utility is allowed to recover its fixed costs from its rate-paying customers. All of these utilities' nonfuel costs are considered "fixed" and are recovered in the base rates charged to the captive customers. These base rates include costs to service debt, earn a profit on equity, and pay all labor and overhead for each generating plant. With access to wholesale customers in other utilities' territories, the regulated utilities have bid the selling price for wholesale power down until it just covers the extra fuel cost they need to run one of their inefficient but underutilized power plants. Since all of the utility's fixed costs are covered in the base rates charged to the captive customers in their monopoly territory, it makes economic sense for them to run underutilized power plants whenever the selling price exceeds the marginal cost, i.e., the cost of added fuel.

This has caused wholesale electric rates to plummet. There were times in the 1996–97 period when wholesale power was sold from hydroelectric facilities as low as 0.8 cents per kilowatt-hour, and wholesale power regularly sells for 1.4 to 1.8 cents per kilowatt-hour during non-peak load periods. These prices cover only the fuel cost and this has been a boon to the 2,014 municipal electric companies in the United States who can now shop the market for power and avoid paying for capacity, labor, or capital recovery. The municipals can buy power at a price that only covers the fuel costs. Only 625 of these municipals, or roughly one-third, own any generating assets, so wholesale power competition is a big help to municipals. Consumers served by the monopoly pay for the fixed costs but they are not allowed to shop for their power. This law turns out to be a subsidy to municipal power companies, and numerous cities are now considering a process to condemn the local electric distribution networks owned by private companies, set up a distribution-only municipal utility, and take advantage of this quirk in the law. Another solution would be to let everyone shop for power.

This opening of wholesale markets without ending monopoly has not done enough to ensure needed capital turnover. The 1992 EPACT has also not helped power entrepreneurs trying to justify building new efficient combined heat-and-power plants to replace inefficient old plants. These power entrepreneurs are at a big disadvantage. The monopoly utilities are already getting paid for their capital, labor and overhead, and they offer power at a price that just covers fuel. A new non-utility power plant has no "rate base" to charge all their capital and labor. The non-utility power plant is only allowed to sell power at wholesale. There simply is not enough gain from heightened efficiency to cover the cost of a new plant, its overhead, and profits by selling power at just the cost of fuel in the inefficient plant. Thus, efficient plants are not being built to replace old plants to any great extent. Society and the environment are losing. Allow these non-utility plants to sell their electricity at retail and they will rapidly replace less efficient older plants.

Retail prices have averaged 6 to 8 cents per kilowatt-hour but have ranged up to 17 cents per kWh in Long Island, New York. If a combined heat-and-power plant were allowed to sell its electricity at retail prices, then such plants would make economic sense and would be built and society would begin to retire the old and inefficient plants. This ability to sell electricity at retail to anyone is critical, because the heat and electric loads do not "match" on most sites. By this we mean that the heat requirements will sometimes exceed the heat available from just generating the electric demand on site. In this case, the plant needs to sell electricity off-site to optimize efficiency. A new combined heat-and-power plant with the right amount of waste heat to satisfy the host usually produces too much or too little electricity. If that plant can sell the excess electricity to others at retail, then combined heat and power plants will make economic sense. In fact, Massachusetts passed a law in late 1997 to deregulate electricity and open retail markets on March 1, 1998. All customers have to pay a "nonbypassable" transition charge for five to eight years to support the payments to the old monopolies for their stranded costs. However, power from a combined heat-and-power plant with over 50 percent efficiency is exempt from the transition charge. Massachusetts will provide a fascinating laboratory for the ideas of this book on promoting efficiency.

ACTION: As described above, we need to open the retail markets to competition and let every customer choose his/her power supplier.

Laws Ban Transmission Competition

There is an almost universal ban on building new transmission or distribution wires by any company except the monopoly utility. The movement of electricity down a street was seen as a natural monopoly, as described in chapter 2, and governments thought society would benefit from monopoly protection of distribution. Some states did not ban municipal power companies from running wires in competition with investor-owned utilities, and some thirty cities have dual wires. There is no evidence that this raises the costs that consumers pay for power. The exception which proves the rule is in Colorado, where the state supreme court has upheld a case in which a person or firm that constructed its own natural gas pipes was allowed to cross streets and serve others as long as it did not offer services to the general public. It appears the case could apply to electric wires as well. This seems an elegantly simple way to induce some competition.

The rules forbidding anyone but the utility to install wires were created to induce fledgling electric companies to build infrastructure—to electrify society. Society achieved that goal four decades ago, but we still forbid competitors to string a second wire. Even with this proposed law change, very few new wires are likely to be strung. Instead, the local distribution utility will negotiate a lower price to keep some of the fees they now charge for transmission or distribution of electricity through their wires. Since the local utility currently maintains the wires, they will usually be able to offer a lower price for moving electricity than the cost of a new wire. What will happen, however, is to let market forces drive the prices of distribution and transmission to the true retail value of moving electricity across the street.

Several state deregulation plans call for an independent system operator to provide nondiscriminatory access to its wires. This will level the playing field, so all generators pay the same price, but it leaves distribution a regulated monopoly with no competitive pressures to reduce costs. Independent system operators will not drive down the price of transmission and distribution without competition from potential new wires.

ACTION #2: We should allow any person or firm to run new wires to supply neighbors with electricity.

Interconnection Rules Controlled by Competitor

It is vital that any power plant operates in parallel with the grid. This takes some explanation. Electricity-distribution systems use alternating current. Motors and generators depend on this alternating current to run, and once they are operating, have a tremendous inertia. Imagine holding two dumbbells at arm's length and spinning as fast as you can. Now imagine trying to stop suddenly and you can gain some sense for the inertia of a spinning generator. They are, after all, large masses of spinning iron and copper, usually rotating at 1,800 to 3,600 revolutions per minute. The electricity from an on-site power plant can be controlled to be identical to electricity on the grid except for the exact time at which the current alternates from positive to negative and back to positive. Electricity in the United States alternates 60 times per second from negative to positive and back to negative. All of the motors operating on this electricity develop strong fields and inertia. If that plant stops producing electricity for any reason, it must shut down for several seconds while the electric fields decay before it can safely reenergize with power from the grid. The current alternation on the grid may not be synchronized with the current from the on-site plant. Alternately, the on-site power plant can be operated in parallel with the grid such that the alternation from positive to negative is precisely synchronized. In this case, grid power can take over instantly because it is "in phase."

This is a very serious issue. The more the on-site plant is out of phase, or sending positive voltage when the grid is sending negative voltage, the larger the problem of an instant transfer to the grid's power. The grid, connected as it is to many generating stations, effectively has infinite power to stay on the same frequency and phase positioning. Connecting the grid to a spinning motor or generator that is out of phase has caused explosions and extensive damage. In the event the on-site plant has an outage, failing to instantaneously switch to the power from the grid is an equally serious problem. All of the user's devices that supply volatile computer memory require uninterrupted power to maintain that memory. A power interruption of just one cycle or one-sixtieth of a second will cause memory loss by computer chips. Factories, universities, medical centers, casinos, prisons, office buildings, and even residences can face significant problems due to power interruptions. Every reader will have experienced minor problems in the home after any power interruption. You know because you must reset all of your clocks, coffeemakers, and VCRs.

The problem is more critical for a factory: A split-second power interruption causes entire manufacturing lines to stop and they must often be reset, sometimes at a cost of hours or even days of downtime. An early customer that Trigen served in Queens, New York, was Seal Kap Packaging, which made waxed-paper cups for yogurt. The plant had a computer-controlled system of pipes filled with hot wax. A short power outage put everyone in a frenzy to reset the computers, because after an hour or so of no control, the melted wax coursing through all of the supply pipes would cool and solidify. If this happened, cup production would stop for three weeks while workers used blowtorches to melt all the wax out of the line. For a hospital, critical heart monitors lose their memories. Clearly, there is a high value in avoiding even one cycle of power interruption.

An on-site combined heat-and-power plant (indeed all power plants) will stop running from time to time, and this can pose a problem for customers unless the plant operates in parallel with the grid. It is vital that the switch from on-site plant to grid, or vice versa, is without interruption. This is exactly the way the grid operates between electric companies and between all large, monopoly-owned plants. When any plant on the grid stops operating, other plants automatically supply more electricity to make up the loss.

This technical explanation was necessary to clarify what has been and continues to be a major barrier to efficient power generation. In every state, the monopoly-protected power companies are given the right to determine what is a safe and proper interconnection. Although this may seem logical, the unintended consequence is to give the monopoly holder a powerful way to discourage efficient on-site plants. Before PURPA mandated that all utilities allow a qualified cogenerator to connect to the grid in parallel, the monopoly utility simply refused to interconnect in parallel, or sometimes at all. As a result, our first cogeneration plants in New York City in 1978 were "stand alone" (i.e., not connected to the grid).

Let no one think monopoly utilities are slow and lumbering in all things. When it comes to preserving the monopoly, they have shown ability to act fast and savagely. In 1977 the Cummins Cogeneration Company (I was the general manager) sold Curt Beusman an on-site cogeneration plant to power his recreation facility, Saw Mill River Courts. The facility has indoor tennis and racquetball courts, a swimming pool, and a health club. We installed two Cummins diesel-engine generator sets with heat recovery to heat the facility and the showers. Either of the two generators could carry the full load at peak, so that the plant had 100 percent internal redundancy. At 10:40 P.M. one night in 1978, the new power plant was able to generate for the first time, and opened the main breaker, thus stopping the current flow from Consolidated Edison. At 8 A.M. the next morning, a Con Edison crew arrived with a crane and flatbed trailer and removed the Con Ed transformer so any restoration of power would take weeks. I do not like to guess at others' motives but would note that if Con Edison hoped our plant would fail, this was a good way to make an example to others who were considering on-site power generation. The message was clear—don't mess with your local monopoly utility.

PURPA eased, but did not eliminate this barrier. PURPA required utilities to interconnect and operate in parallel electrically with a "qualified cogenerator." If your plant qualified by being 50 percent more efficient than the national average utility generation, the local monopoly had to interconnect in parallel.[107]

Post-PURPA, Con Edison would be required to offer backup power to Saw Mill River Courts. However, monopolists still have the authority to approve or deny the interconnection design. PURPA changed the game only slightly. Forced to parallel, the monopolist's next anticompetitive strategy was to drive interconnect costs as high as possible.

Approval of interconnection can also be a time barrier. Colorado law allows the utility five full months to review the interconnect design, and if any changes are required, the five-month clock starts over. In this case, the competitor can stall a combined heat-and-power plant design for at least ten months, if not more. Regulatory commissions fear taking responsibility for the safety of the electric system and monopolists skillfully exploit this fear. In addition to stall tactics, most utilities require a needlessly expensive interconnection that can cost three to four times as much as a simple and functional but safe interconnection. In fact, utilities start with a seventy-year bias in favor of overinvesting (because that is how they increase profits under regulation) and often don't realize how unnecessarily expensive their own interconnections are. By demanding overinvestment in interconnection and then delaying approvals, the monopolist can often frustrate construction of a competing small generation plant.

One of the more egregious examples of interconnection overdesign we have seen is in Canada at Trigen's small, 4-megawatt combined heat-and-power plant that supplies steam to downtown London, Ontario. Ontario Hydro, the provincial monopoly, forced Trigen to remove a newly built chain-link fence surrounding the property and replace it with an off-standard, specially manufactured chain-link fence of their design. Next, Trigen had to ground the fence with copper wires, although the fence was never closer than twenty feet to the generator, and the generator was inside a building. Trigen even had to ground all of the steel pillars in the building. This increased the price of the plant by 5 percent for no technical or safety reasons whatsoever. In fact, our plant is connected to miles of steel pipes that are buried beneath the streets, and form the most elaborate possible grounding system. Ontario Hydro's issue was not safety. They used their regulatory powers to make competition more difficult. When we began to operate and sell power to London Hydro, the municipal distribution utility, Ontario Hydro brought a lawsuit (still pending) to unseat the London Hydro commissioners because they had not obtained Ontario Hydro's permission to sign the contract. Meanwhile, Ontario Hydro announced the permanent shutdown of 4,600 megawatts of its own nuclear plants for safety reasons. Yet they fight our small efficient 4-megawatt plant.

ACTION: The DOE or some other nonpartisan group should develop national standards that apply to all interconnections, removing this authority from all monopoly utilities. The body setting the interconnection standards should be independent, and review improvements in technology periodically to update their rules to allow modern approaches. The DOE's goal should be to minimize the cost and approval time for safe interconnection.

Backup Power from Competitor

All machines and all power plants need some downtime for maintenance, and all machines will break and become inoperable from time to time. To provide reliable electric power, these plants must have a source of backup power. Before PURPA was enacted in 1978, utilities could refuse to provide backup to any on-site plant, so the only way to achieve reliability was to spend more money and build redundant capacity, as we described above for Saw Mill River Courts, where we installed 100 percent redundancy. Except for very high-cost electricity areas, like the New York City area, the cost of this redundant capacity usually made the on-site plant uneconomic. This was an effective deterrent to the construction of on-site plants. PURPA mandated that the utilities provide backup power for qualified cogenerators at fair rates. But what is a fair rate and who decides what a fair rate might be?

If many relatively small plants purchased backup capacity from the central utility, all of them would not be expected to fail at the same time, so the total backup power drawn at any given time would be only 2 to 5 percent of the total contracted backup power. However, utilities have argued to their regulatory commissions, usually with some success, that a competing plant would or could fail at precisely the time the utility system was experiencing a peak electric load. If you accept this argument, then the backup price should reflect the full cost to the utility of adding the exact same amount of new generating capacity. The first big problem with the argument is that monopoly utilities have installed as much as 25 percent spare capacity for reliability and this capacity is already in the rate base, i.e., already being paid for by the monopoly customers. Clearly, the utility would not build more capacity to meet a small backup obligation, and any money it receives is new revenue. But utilities do not want to lose the customer and have logical reasons to make on-site generation uneconomic with high backup charges.

Perhaps you are thinking, "Wait a minute. Why should the utility dedicate any of its spare capacity to a competitor, even if it has a large reserve margin? Won't the utility need to maintain the reserve margin and have to build new generation anyway?" To start with, the utility had set its reserve margin with the customer's load included. If an on-site plant generates the customer's load most of the time, that customer's demand is no longer part of the utility's peak demand, so spare capacity exists. The situation is more complex for the real world where all utility plants are interconnected with other utility plants in large grids that span half of the United States. When there is a need for extra power anywhere on the grid, any generating plant that is connected can increase its output and satisfy the extra demand. In fact, this is automatically done every minute of the day. It is not clear that the utility that has been asked for backup power will ever have to build new capacity to be able to offer emergency power.

If a company sells one life-insurance policy that will pay $1,000 in the event the insured person dies, there is an exposure of $1,000. If that insurance company sells 10,000 such policies, it is inconceivable that all 10,000 of the policyholders will die at the same time; thus the financial exposure of the insurer is subject to actuarial probability. Actuaries review mortality rates and ages of the insured, and then predict with great confidence the probable exposure to the insurance company in any given year. This leads to insurance-policy premiums of as little as $20 per year for $1,000 coverage. The same logic is true for backup power, but monopoly utilities have claimed they must maintain 100 percent backup, i.e., charge the equivalent of $1,000 per year for a $1,000 life-insurance policy. This blocks self-generation.

Monopolists are not going to help a competitor voluntarily by using actuarial methods to determine how to lower the backup rates for combined heat-and-power plants, since they would prefer to keep the existing customer to themselves. It has been largely beyond the capability and interest of most state utility commissions to deal effectively with this issue. If the commissions were confronted with a large organized group of would-be cogenerators, those commissions would probably act. The reality has been that the potential cogenerators do not even realize they have an option. Each David appears alone to fight Goliath, and Goliath wins most of the time. The result has been that backup power rates are a very real barrier to the widespread building of efficient power plants.

The backup power technology indeed is very common. Utilities have purchased power from each other for more than seventy years on a spot basis, hour to hour. With the exception of very rare major transmission system failures, utilities have always been able to buy backup power. The three-hour electrical outage experienced by the average American consumer each year is largely the result of local lines being downed by storms. Power outages are almost never a result of adequate generation in the system. What societal purpose is served by denying this always available backup power to the combined heat and power plant?

The power plant serving Coors Brewery Company in Golden, Colorado, that was purchased in 1995 by Trigen Energy Corporation and our partner Nation's Energy, a subsidiary of UniSource, is an example of how the present monopoly of backup-power supply damages economics. Unisource's regulated subsidiary, Tucson Electric Power, is part of the regional pool of monopoly utilities that regularly exchange electric power with Public Service of Colorado. The transmission wires to move the power from a Unisource generation plant to Golden, Colorado, not only exist, they are used to back up Public Service of Colorado. However, our very efficient cogeneration plant that supplies Coors with both steam and electric power is forbidden by law to purchase backup power from anyone but the local monopoly utility, Public Service of Colorado. As a result, we pay significantly higher prices for backup power compared to what we would be charged by our partner, Tucson Electric Power. In fact, we estimate that the $2 million we pay each year for having backup power available is roughly four times the market price.

ACTION: It should be mandated that non-utility power plants have the right to purchase backup power from any generator and that the local utility must transmit the power at a price no higher than what they charge other customers per kilowatt-hour. Let the market decide what the likelihood of a power outage would be and whether that outage will come at a time of system peak. Then, let the market set the prices for backup power. Let whichever generator has surplus capacity bid to sell backup power to any other generator, hour by hour.

Condoned Anticompetitive Practices of Monopolists

The Sherman Antitrust Act and the subsequent Clayton Antitrust Act were established to ensure that large and powerful firms would not use their financial strength and market power to force competitors out of business. These acts were also designed to prevent firms from bundling products to force a supplier of only one product out of the market, and they prohibit all firms from selling below cost to force someone out of business. However, by and large, monopoly utilities have been allowed to function outside the antitrust rules. The unintended consequence has been that monopolists have employed precisely the tactics that these antitrust acts bar to prevent competition from other products like heat and chilling. The following are some of the problems that have resulted from condoning antitrust actions. Such practices are effective barriers to combined heat-and-power plant development.

Bundling Heat and Power

In most of the United States, the utility's electric load peaks in the summer, due to heavy air-conditioning demands. There is a lower system peak in the winter. It is thus logical for the utility to encourage customers to use more electricity in the winter, i.e., for heat. Utilities increase the demand for electricity in the winter by selling electric heating and thus putting to work idle capacity that was built to serve summer loads. To increase the demand for electric heating, the utility might lower the price it charges for electricity used for heat closer to its marginal cost of production. Nothing in the antitrust law prevents this pricing strategy, but under current regulations, the utility would then have to offer the same price to all similar customers, and this would severely depress the utility's revenue.[108]

What has happened instead, fully condoned by the regulators, is that users are given lower electric rates if they agree to build and operate all-electric buildings. These special low rates apply only if all of the building's heating, cooling, and other energy needs are provided by electricity. In other words, users receive a lower price if they purchase the entire bundle of services from the electric utility—if they use electricity for all their energy needs. If a building uses package boilers for heat or gas-fired chillers for cooling, it loses its bargain electric rate. Since the local steam-system operator is forbidden to compete with this bundling of services by offering electricity, the playing field is not level. Neither is the local gas or oil company (who offers alternatives for heat) allowed to compete and offer electricity. All-electric bundling squeezes out the gas and oil companies and the district heating and cooling company. All-electric rates have been a very effective way to keep buildings from buying heat from steam systems, or from generating heat themselves. Burning three megawatt-hours of fuel to make one megawatt-hour of electricity, and then using that electricity to power a resistance heater that produces one megawatt-hour of heat is an extremely inefficient way to produce heat. The same heat could be made in a boiler with only 1.2 to 1.3 megawatt-hours of fuel. Electric heat uses more than twice as much fuel as producing heat in a boiler, and sometimes displaces waste heat available from a cogeneration plant. By and large, electric heat could not compete with gas-fired heat or with district energy, but for the antitrust practice of bundling all products. This bundling prevents competition for heat and creates needless burning of fossil fuel.

ACTION: The Federal Justice Department should aggressively apply antitrust rules to all monopolies. The government should make it illegal for monopolists to bundle electricity so they can no longer use this technique to force other heating and cooling suppliers out of the market.

Buying Out Cogenerators

The only way around bundling for a would-be competitor is to build or have built an on-site power plant, usually a cogeneration plant. This plant may achieve up to 90 percent efficiency by recovering normally wasted heat-thermal energy to heat and cool the building. In most states, if the electricity produced by the plant does not cross a public street, it is legal to sell that electricity and the heat to the owner of the building or factory. This does not violate the monopolist's franchise. As cogeneration developers have tried to offer savings (and save fuel) with on-site plants, monopolists have acted to prevent any demonstration that on-site plants work. After all, if several combined heat-and-power plants were successfully built and operated, other users and the regulators might begin to develop doubts about the monopoly paradigm. The monopolist's first line of defense is to do all of the actions discussed above to discourage the plants from being built—high interconnection costs, delays in design approval, high backup charges, bundling of products to shut out the sale of heat, and even refusal to parallel with the on-site plants. As the technology has improved and independent power companies have learned how to counter many of these arguments, occasionally these defenses are breached and a user decides to purchase energy from a combined heat-and-power plant.

An increasingly common practice of monopoly utilities has been to wait until customers receive a bona fide offer for lower-priced energy from a developer of combined heat-and-power plants, and then offer a secret deal to buy out the proposed on-site power plant and maintain the electric load. The monopolist's argument to the regulatory commissions is as follows: The customer that is threatening to exit the system is paying part of the overhead and capital costs of the monopoly utility. If that customer leaves the system, then the remaining customers will be forced to pay higher rates (by perverse monopoly logic, certainly not true for a competitive business). Thus, the monopolist argues, it will minimize the damage and avoid an increase in rates to other customers if the utility offers to give up part of its margin to hold the customer who is considering an alternate supplier. All is done in secret. Why? Maybe because all other customers would demand the same treatment? Maybe because the rate offered is actually below cost?

Trigen has on several occasions worked with customers for months and been close to signed agreements to proceed with a new combined heat-and-power plant. Then the local utility has offered a huge discount ($20 million in one case) to retain the customer, each time with a secret deal.

In Atlantic City, New Jersey, casinos are ideal candidates for combined heat and power. Each has a large hotel attached, and each operates with near twenty-four-hour-per-day activity on the gaming floors. Casinos have a steady demand for lighting, and hot water for restaurants and showers. Trigen developed proposals to several casinos that would cut their fuel use in half and provide in excess of $1 million per year in savings as well as provide a return on our investment. In two cases, casinos tentatively accepted our proposal and were moving toward contracts when the local monopolist, Atlantic Electric, offered secret deals with lower rates that matched or beat the economics Trigen was offering. In both cases, the New Jersey Board of Public Utilities approved the secret deals.

If this is good public policy, why does the decision review exclude one of the impacted parties? Antitrust laws consider a market-dominating firm to be dumping its product when it sells below cost. These antitrust rules are not limited to some portion of the cost, but to all of the cost. Yet, utilities are, in some cases, allowed to block efficient combined heat-and-power plants by selling below cost. In this example, the New Jersey Board of Public Utilities did not seem to consider the values of increasing efficiency or reducing air pollution, but no one really knows since the hearings were held in secret. This is not usually the way of democracy.

ACTION: There should be public hearings on all deals offered by monopolists until the monopoly protection is ended. Secret deals have no place in the electric marketplace. Allow the existing antitrust rules to govern market behavior.

Real-Time Pricing

A recent technique for encouraging customers to balance the system loads of a central utility is to offer "real-time pricing" (RTP) to those who have the ability to switch loads off or to backup generators during the utility's peak hours. Under RTP, the customer pays a different price every hour, based on what the marginal costs of production are for that hour. We use such pricing in Tulsa, Oklahoma, where our plant can produce chilled water with either electric or steam-driven chillers. Whenever the RTP electric price is low, we use our electric chillers. During hot days, the load on the Public Service of Oklahoma system grows and they must operate their least efficient generation plants to meet the electric load. The marginal cost of electricity during these hours is very high, so we switch to steam-driven chilling. This actually helps the utility level its load.

However, in a comparable city in the same region—the local electric utility—offered RTP (under a pilot program) to the local convention center—a Trigen chilled-water customer. The city fathers were assured that they would be able to purchase cheap electricity, but their convention center has no ability to curtail loads at peak hours. A convention center is an illogical candidate for real-time pricing, because prices for electricity are highest at precisely the times a convention center needs chilling. At night when the system loads are low, the utility charges low rates under real-time pricing, but the convention center is closed. It made no sense that the convention center should be singled out for special prices given their very poor load factor. (Convention centers only operate when there is a trade show or convention, and are often empty.) When we challenged the proposal, the state commission said the real-time prices were secret and the utility did not have to make them public. On the strength of the utility's oral pledges, the city decided to spend several million dollars and build its own electric-driven chilling plant and leave our district cooling system. We have calculated that the actual cost of electricity to power the electric chillers will be more than double the oral promise, and we believe the city will lose, but we were unable to challenge the pricing of this secret deal. This approach permits a monopoly to misapply a rate like real-time pricing and to cross-subsidize one customer with money charged another, and thus use below-cost rates to block competition.

Perhaps more damaging, the utility refused to sell our chilling plant RTP, even though we are an ideal candidate, because we can switch to electricity when supplies are abundant. Denying us RTP rates prevents us from lowering our cost of chilled-water production.

ACTION: Apply existing antitrust rules to electric utilities.

Percent of Load Rate Ratchets

Another effective way to retain most of the large customer's load is to apply penalty rates when that customer "outsources" too much of his total electric load. This type of rate schedule is common in industrial states like Illinois and Iowa, where some cogeneration has always been practiced by large industrials. One Iowa utility has an approved rate structure that says if the percent of load purchased by any customer drops below 50 percent of that customer's total electric demand, then the charge per kWh for the remaining electricity nearly doubles. In other words, if customers generate on-site more than half of their electric needs, they better generate 100 percent, because the penalty rate from the monopolist on the remainder will result in much higher total costs. The penalty rates are higher than would be paid by a smaller customer who had the identical load, but purchased their entire load from the monopoly utility. How would we respond to a supermarket that doubled the price of our groceries if we did not purchase the majority of our food from its store? Mercifully, supermarkets do not enjoy monopoly protection.

ACTION: Existing statutes should be applied. The practice of penalizing customers who purchase less than all of their needs is not allowed under antitrust statutes, but is common practice for monopolies.

Environmental Laws Ignore Efficiency or Assume Central Generation

The Clean Air Act of 1972 was passed to improve air quality, and it targeted electric power plants, given the amount of pollution they produce in a single location. Assuming all power plants are large, the law regulates all new power plants. The primary focus for the clean-air regulations has been to mandate that new power plants use state-of-the-art pollution controls. This sounds like a good approach to cleaner air and has not been questioned much, but the approach is flawed. The present approach to environmental regulation has several unintended effects. This approach perversely increases the output of regulated pollutants like nitrous oxide and sulfur oxide to what would happen under smarter rules, as we explain below. The Clean Air Act ignores the best pollution control strategy—burning less fuel. There is no reward for efficiency in the present regulations. There is not even any acknowledgment of efficiency as a control strategy.

Input-Based Rules Ignore Efficiency

The air-emission pollutant allowances for power plants are typically stated in parts per million of pollutant in the exhaust gas. However, what is not understood in formulating these regulations is that the exhaust gas flow from any boiler or power plant is related to how much fuel that plant burns. Burning more fuel produces more exhaust. Why? To combust fuel any fire uses oxygen atoms in the air to join with all of the carbon and hydrogen atoms in the fuel. Air contains about 21 percent oxygen; most of the rest of air is nitrogen. Thus, the total mass of exhaust gas produced by any combustion process will vary up or down in almost precise relationship to how much fuel is burned. If one plant is twice as efficient as another, it will burn only half the fuel, will need only half as much combustion air to supply oxygen, and it will thus have only half the volume of exhaust gas, while producing the same amount of useful energy. This is the definition of efficiency: useful energy produced divided by gross fuel energy burned.

The present environmental rules set NOx standards in parts per million in the exhaust gas, regardless of efficiency. The rules apply equally to a plant with 33 percent efficiency and a plant with 66 percent efficiency, but since the inefficient plant has twice as much exhaust, it is allowed to emit twice as much NOx. Both plants produce the same "parts per million" of NOx in their exhaust, but the inefficient plant burns twice as much fuel, produces twice as much exhaust, and is thus allowed twice as much NOx. Since carbon dioxide emission is a function of fuel burned, the inefficient plant produced twice as much carbon dioxide as well. This approach is termed an *input-based standard*.

What is tragic is that we could just as easily measure and administer an output-based standard. The EPA could rescind present input-based standards and instead set performance or output-based standards for each regulated pollutant. For example, the EPA could allow every power plant "X" pounds of NOx per megawatt-hour of electricity produced, and "Y" pounds of NOx per megawatt-hour of heat produced and used. This output-based standard would encourage every designer, owner, and operator of a power plant to not only consider all technologies that remove NOx from the exhaust, but to also consider a new pollution control strategy—efficiency.

The Clean Air Act's current approach actually increases carbon dioxide. Pollution-control devices are nearly all electric. Adding such a device worsens a power plant's fuel efficiency and thus causes an increase in carbon dioxide emissions. With input-based regulation, the only way a power plant can produce fewer parts per million of NOx, SOx, and particulate pollution is to apply energy-consuming pollution-control devices. With performance or output-based regulation, power plants could also comply with EPA goals for criteria pollutants by recovering waste heat to sell to nearby industrial concerns, medical centers, or universities, thus gaining credits for NOx per megawatt-hour of heat recovered. Since the nearby facilities would shut down their boilers, carbon dioxide, NOx, and other pollutant output would drop. Alternately, the market could seek the cheapest way to reduce NOx and then trade permits. Output-based regulation of NOx, SOx particulates, and other pollutants will automatically encourage the entire industry to conserve energy. This will lower carbon dioxide and save money.

ACTION: The EPA should replace all input-based standards with performance or output-based standards that set pollution allowances per unit of heat and per unit of electricity produced. These should apply to all power and boiler plants.

The Present Regulatory Approach Extends Use of Inefficient Plants

The EPA's approach to reducing pollution has been to require all new plants to meet New Source Performance Standards (NSPS), which are constantly made more severe as the technology of pollution control improves. In areas that do not currently achieve the National Ambient Air Quality Standard, the emission limits are significantly lower for new equipment. In these nonattainment areas, emission limits are set equivalent to the Lowest Achievable Emission Rate (LAER). This approach clearly induces a race among engineering firms to develop new and better control technologies, which seems good. Once demonstrated to work, any new, improved pollution-control device finds a ready market. California, New Jersey, and Massachusetts, which are filled with nonattainment areas, and other states quickly mandate all new power plants to use the new control technology, creating an instant market for the new pollution-control device.

This sounds like a logical way to clean up the environment, but is a major reason we have so many old, dirty plants still in operation. A new gas turbine, without any added controls, emits about 10 percent of what the oldest power plants emit. The states that do not achieve ambient air-quality standards force a would-be builder of a new plant to add pollution-control devices that increase capital costs by up to 20 percent and use up to 5 percent of the power to operate the pollution-control devices. All of these added controls may cut the gas turbine's pollution in half or to 5 percent of the pollution from old power plants. This slows the replacement of old plants. These states need to employ the "Willie Sutton" principal. The judge asked Sutton, a real-life thief, "Why do you rob banks?" Sutton replied, "That's where the money is." The pollution "money" is in old plants. The unintended consequence of exempting pre-Clean Air Act plants from emission regulations has been to increase greenhouse gases by preserving the stock of existing, inefficient, and dirty power plants.

There are thirty- to forty-year-old peak shaving combustion turbines along the New Jersey Turnpike and throughout the country that emit over 600 parts per million of NOx in a visible cloud of yellow exhaust. They have roughly 20 percent efficiency and burn expensive jet fuel or kerosene. These emissions are allowed because these turbines are "grandfathered." They were installed before the Clean Air Act. Fortunately, these turbines are only used for peak power demands and do not run very often. Drive the New Jersey Turnpike in July on a hot, muggy day and you will see the yellow cloud of polluting NOx.

The many very inefficient gas-turbine peakers that are allowed to operate without pollution controls contribute to carbon dioxide and dirty air, but since they only operate a few hours per year, the problem is small. Their relatively cleaner brothers are the real problem. Many old base-load power plants—plants that run all or most of the time—have NOx emissions in the range of 200 to 300 parts per million of exhaust, and achieve less than 30 percent delivered efficiency, but they are also "grandfathered." They run around the clock and add significantly to total air pollution in the United States. By contrast, some current state regulations, notably in California and Massachusetts, demand that all new power plants, of almost any size, have NOx emissions as low as 6 to 9 parts per million. Now adjust these figures for useful output of electricity. Since the new combined heat-and-power plants have double the efficiency of the older grandfathered plants, they have only half as much exhaust per kilowatt-hour. This means the maximum emissions allowed from new plants are equivalent to 3 to 5 ppm from the old plants. In other words, in certain states it is illegal to replace generation from a plant with 200 ppm of pollutant in its exhaust with a new plant that emits above 5 ppm equivalent. The result is to keep the old plant in service and build new plants only for increased load. Surely we can make better rules!

This approach to lower overall emissions by constantly lowering allowed pollution from new plants in nonattainment areas is a barrier to efficiency and to clean air. This approach has worked with automobiles because cars wear out and are replaced in about ten years. Power plants don't wear out. Operators can repair them for fifty to seventy years and do so, largely due to Clean Air Act methodology. The current regulatory approach reduces the attractiveness of building a new generation plant, so to satisfy the demand for power, old, inefficient, and dirty plants keep operating. They are not subject to the regulations.

Over the past twenty years, the regulations have been tightened ahead of the development of economic-control technology and have become a nearly insurmountable impediment to building small combined heat-and-power plants. For larger central plants, very low pollution has a somewhat lower cost per kilowatt of capacity, due to economy of scale, but still serves as a major disincentive to build new and efficient plants.

The facts are quite clear. Over two-thirds of all generating capacity in the United States was built before the Clean Air Act in 1972 and is "grandfathered." It is often cheaper to extend the life of these twenty-five-year-old and older plants than to build new doubly efficient plants that are forced to use the best available control technology to emit 5 percent of what they replace instead of 10 percent of what they replace. This approach keeps entrepreneurs from building efficient new plants that would reduce pollution by 90 percent.

Recently, the EPA has proposed particulate reductions by the old power plants, based on expected health benefits of fewer particulate emissions. These new regulations would set particulate standards for many existing plants and would significantly increase their costs. A better approach to lower particulates is to encourage a turnover of old generating plants to more efficient generation. The proposed new particulate regulations follow the hated "command-and-control" approach. Switching to output-based standards for currently regulated pollutants which apply to all power plants would let the market find the cheapest way to lower total emissions, and would change the present bias *against* new plants to a bias *for* new plants. The EPA could simply calculate the average U.S.-wide production of NOx per megawatt-hour of electricity and per megawatt-hour of heat produced, and then mandate that every power-plant operator meet this standard. These EPA regulations could ratchet the NOx standard per megawatt-hour down over time to meet air quality goals. The new regulations could allow trading to achieve emissions targets and avoid fines. In other words, this would let the market work (chapter 9 deals in greater depth with this proposed approach as a way to control carbon dioxide).

This change to EPA regulations would create a boom in new, higher efficiency combined heat-and-power plants near heat loads and would clean the air. These new, highly efficient plants, even without expensive add-on emission controls, will achieve a tremendous reduction in pollution, and are doubly efficient. Carbon dioxide emissions will drop as we save money. Utilities, seeking to meet the tightening standard of average pollutant/megawatt-hour, would study the cost of new pollution-abatement controls, the cost of operating old inefficient plants, and the cost of building new, more efficient plants. Under enlightened rules, utilities would often conclude that it makes no economic sense to continue to extend the life of dirty, old, inefficient power plants.

ACTION: The EPA should eliminate rules that require the Lowest Achievable Emission Rate (LAER) for new power plants and replace them with output-based standards that are applicable to every existing and new power plant, and then establish a pollutant-trading program to allow the market to achieve the cheapest pollution reductions. The standards for each pollutant should be lowered over time, based on a schedule.

This change of regulatory approach would ease the workload on each state regulatory office and greatly lower the friction created by the present approach. Adopting output-based standards for all power plants would take the EPA out of the present business of dictating specific control technology and would end the controversy about how soon any specific technology is ready to be deployed commercially. The EPA would instead focus on defining the present and future standards for acceptable levels of NOx and other pollutants per megawatt-hour of heat and electricity produced and would leave to the market the task of reaching these standard targets. This change would achieve more pollution reduction at lower cost than the present approach.

Air Permits Required to Start Construction Delay Efficient Plants

In 1972, when the Clean Air Act was passed, pollution control was new and hotly debated. The lawmakers at the time apparently assumed that utilities and industrialists would try to avoid compliance. There was a fear that if a new power plant started construction before receiving an air permit, the plant owner would gain political leverage to receive a permit even though the plant did not ultimately meet the emission requirements. The fear was one of special pleadings that said, "Look, we spent all this money and did our best, so even though we did not reduce emissions as much as you thought possible, you must give us a permit." To avoid this chance for special pleading, the original Clean Air Act of 1972 forbade starting plant construction until an air permit was granted.

This is not the usual permitting approach. Commercial office buildings gain zoning approval, and then begin construction before receiving complete building-code approval. Developers know they must comply with the building codes and that occupancy will not be allowed until the authorities grant a certificate of occupancy, but developers are allowed to proceed with construction. During construction, building inspectors examine the work and may mandate changes before issuing the Certificate of Occupancy. The developer risks that his or her team understands and will meet the codes and builds accordingly. Delays are avoided. Since commercial buildings are not monopoly protected, the risk of not meeting codes and not being allowed to rent to tenants is the developer's risk. This is not the case with environmental permits.

If any firm begins construction of a new power plant before the air permit is issued, that firm is subject to fines and the CEO of the firm is subject to a jail term—serious penalties indeed. Obtaining the air permit from overworked and understaffed departments of environmental protection often takes eighteen months or longer, so construction is delayed. Yet, many of these plants use standard gas turbines or diesel engines and use well-tested, well-known pollution-control technology. The regulatory approach is rooted in the idea that all power plants are large and unique central plants. The drafters of the law and regulations simply assumed that all new power plants would be giant, uniquely designed, site-constructed, electric-only plants—just what monopoly utilities always built. The drafters of the Clean Air Act assumed each power plant would be so unique as to require detailed study by the permitting agency. They were right to fear that a large monopoly utility, after spending billions on a new plant, would have political leverage to gain an air permit even if the plant was out of specification. This approach makes no sense for small combined heat-and-power plants that use mass-produced engines, but the law applies to all power plants, regardless of size.

Paradoxically, anyone can purchase and operate a new engine or gas turbine without an air permit, providing it is powering a truck, car, airplane, bus, or a portable generator set. The manufacturer of the engine meets EPA standards once for each model, then certifies that every engine it builds will have the same emissions and meet the rules. If any developer wants to use the identical gas turbine for making electricity at a fixed location, he or she must explain every aspect of the technology to an environmental permitting agency, who then carefully studies the application and takes up to eighteen months to issue an air permit. This regulatory process slows the deployment of efficient combined heat-and-power plants and often completely discourages their construction.

ACTION: All power plant developers should be permitted to build, at their risk, without air permits, but then not be allowed to operate their plants until they comply with air-quality rules. Standardized permit conditions should be established for all mass-produced engines and boilers. Regulatory approaches to promote rapid deployment of more efficient plants should be streamlined.

Businesses should decide what control technologies to deploy and where to deploy them. Let those businesses take the risk of the technology working well enough to meet air-quality rules. Again, regulate those plants with performance standards based on the output of pollutants per unit of useful energy produced.

Allow Pollution Offsets

A new combined heat-and-power plant usually supplies heat to existing buildings that have preexisting heat-only boilers. If a new cogeneration plant is built, the heat it recovers reduces the heat needed from the existing boilers and lowers emissions from existing boilers. A new combined heat-and-power plant using today's superclean gas turbines can have a dramatic impact on air quality, especially local air quality. The heat-producing boilers that are relegated to backup service are usually unregulated, lack pollution controls, and have short stacks, so their pollution stays at ground level. The new combined heat-and-power plant will meet current pollution standards, is very clean, and doubles efficiency.

Now look at permitting rules. When Trigen developed Trenton District Energy Company, we sought emission credits for all the pollution coming from the many boilers we would replace. A federal EPA official told us, "OK, bring me certificates of destruction of the hospital and prison boilers, and I will grant emission credits." Two small problems arose. It was illegal for us to even break ground on our new plant until the air permit was issued. Then we needed eighteen months to build the plant. What was the hospital to use for heat for that year and a half after they destroyed their boilers? Second, it was never our intent or even a good idea to destroy the hospital's boilers. Those boilers are used for backup, in case our hot-water pipes are temporarily damaged. The hospital, Mercer Medical Center, was not going to use any more heat because they kept their old boilers for backup or because they connected to our cogeneration plant. But the rules do not understand this fact. The rules are effective barriers to efficiency and barriers to cleaning up the environment. The EPA rules frustrate the goals of the EPA.

Regulations Prohibit Optimizing Plant Operations with the Grid

This barrier is a powerful example of "the law of unintended consequences," and stems in part from the failure of environmental regulators to understand that power plants do not create demand for power: they merely supply the power demanded by the public. Permits for new sources require continuous compliance with strict emissions controls over the full range of operating conditions for each new facility. The idea of the regulators, which seems logical, is that new sources must operate only in their cleanest range or must shut down. This effectively limits a new, efficient plant to operation in only certain ranges of output. The parts per million of carbon monoxide emissions usually increase by 20 to 50 percent when a gas turbine generates at less than half of its full capacity. This sounds like a big increase in pollution, but remember, the plant was only emitting 5 percent as much CO as comes from most of the electric plants in the country when they generate the same amount of power. With a 50 percent increase in pollution at part load, the new gas turbine plant emits $7\frac{1}{2}$ percent of the pollution of the plant it replaces, but it is required to shut down. The old plant then generates more power and net pollution increases 13- to 14-fold.

Once again, the law of unintended consequences renders this approach counterproductive. The demand for electricity is not subject to regulation. When any consumer on the grid turns on a light bulb, starts a motor, or otherwise uses electricity, some generating plant connected to the grid must generate more power. Without more power, there will be either a brownout, or a blackout for the whole grid. This is an unthinkable consequence, and every grid solves this problem by allowing each plant to automatically increase generation to meet demand.

The utilities try to operate the grid so as to keep their most efficient plants operating every hour. Their older plants are grandfathered and not subject to emission regulations, so these old plants are allowed to operate at part load. A cogeneration plant should not be shut down completely, because it has a steam obligation. Even at part load, the plant is very efficient because it is recovering heat to make the needed steam. However, most new generating plants can't meet the stringent new source emission limits at part-load conditions and thus the rules require them to shut down. These rules can cause a 75 to 90 percent efficient plant that emits only 40 parts per million of NOx to be shut down. Since the power demand must be met, some older plant automatically increases its inefficient production and emits 200 ppm of NOx, all in the name of clean air.

Most of the generation plants connected to the grid were built prior to the Clean Air Act (two-thirds of U.S. power plants fit this category). As noted above, these plants are grandfathered from many air-pollution regulations and they emit between one and two orders of magnitude, or 10 to 100 times, as much pollution as the new plants that meet the latest new source performance standards. The unintended consequence of limiting the operating range of new plants is that the electricity demand is often met by old, dirtier sources.

ACTION: As previously discussed, we need to apply a nationwide standard of pollutants allowed per megawatt-hour for the year to all power plants. Let each power plant operator have an option of buying emission credits from cleaner plants to meet the standard, and then get the regulators out of the way. The market will find the cheapest and lowest cost approach to control pollution every hour of every day. The emissions during an hour when the plant was operating at half load might be higher than when at full load, but the operator would record all data and be forced to meet an average standard for the year for all megawatt-hours produced.

Tax Policy Penalizes Efficiency

The present tax code allows five to seven years of tax depreciation for engines or gas turbines used in airplanes, buses, or trucks. However, if a developer uses the same engine or gas turbine to power a generator, the tax depreciation life is raised to fifteen years. If the developer opts for more efficiency, designs the plant to capture exhaust heat to make steam, and uses the recovered steam in an efficient combined cycle, the tax life is raised to twenty years. We recently asked the Treasury Department to change the tax-depreciation lives on gas turbines and engines to seven years, regardless of their application. The initial negative response of the Treasury was rooted in the old central generation paradigm. The official handling the request noted that utilities had numerous gas turbines in service that were twenty-five to thirty-five years old, so the turbines clearly had a long life and it was appropriate to give them tax depreciation of fifteen to twenty years. However, these utility gas turbines are large, industrial-quality machines that are used only for peak shaving. They only operate 100 to 300 hours per year. After thirty years these peaking gas turbines may have operated the equivalent of one full year. In addition, the monopoly utilities are able to put the turbine in their rate base and be assured of recovering the capital from rate payers.

By contrast, gas turbines and engines used in cogeneration plants run all the time—twenty-four hours per day, seven days per week. In seven years, these turbines will have run the equivalent of seven years—they often operate over 97 percent of the time, and wear out accordingly. The smaller on-site plant depends on the factory it serves staying in business and continuing to purchase power, so the economic life is five to seven years. The tax law, thinking only of central utility plants, treats this small plant's tax life as twenty years.

This tax-code formulation is an example of the impact of a ubiquitous paradigm of central plant electric generation only. Those legislators who wrote the present tax regulations must have thought that all power plants using turbines would use large 150–250 megawatt gas turbines. They simply did not realize they were penalizing smaller plants and efficiency as well.

An interesting side note is to observe how tax depreciation rules are established. For years, this specialized task was the responsibility of Treasury officials who studied actual equipment records and set detailed rules. When Treasury increased the tax life of corporate aircraft, Senator Dole from Kansas intervened and was able to pass a law that requires Congress to set all tax depreciation laws.

ACTION: The tax depreciation life of engines should be based on the type and size of engine used, then applied consistently to every application of a particular engine. Tax credits should be considered for very efficient plants regardless of size. Disincentives that limit the capital-stock turnover of electric production plants so as to encourage more efficiency and lower-cost energy assets should be eliminated.

Federal Reimbursement Policies Do Not Reward Efficiency

The federal government provides extensive financial support for housing, military bases, General Services Administration (GSA)–owned buildings, Veterans Administration hospitals, and more. In almost every case, the funds cover actual expenses; any energy-cost savings, for any reason, reduce the amount of the money paid by the government, dollar for dollar. This sounds totally logical and seems to protect taxpayers. But it has an unintended consequence concerning energy efficiency.

A local housing authority usually manages low-income housing units. The authority limits the rents charged to the low-income tenants to well below the actual cost of the housing. To balance their budgets, the housing authorities rely on local and state subsidies and finally turn to the federal government to make up the remaining difference between cost and revenue. Now watch what happens to a proposal for more efficiency. What happens when power entrepreneurs go to the housing administrator and propose to install an efficient plant, with their capital, and offer to reduce the energy prices paid by the housing unit? In our experience, the result is almost always no action. Why?

The housing authority officials see risks in this proposal, but see no gain if the plant works as proposed. These officials have no incentive to proceed. If the combined heat-and-power plant or other efficiency investment is a success, operates reliably, and saves money, then the federal reimbursement will drop by the exact amount of the savings. The housing officials won't have one extra dollar to spend on things they think the housing unit needs. If the on-site power plant has any problems, the administrator may be considered responsible and may even lose his or her job. The administrator may win accolades for reducing cost but risks losing a job or promotion. This is not a promising bet, and most housing authorities opt for the inefficient status quo. It takes great courage to do otherwise.

We spent eighteen months working through this exact problem in Baltimore. The Cherry Hill Housing Project includes 1,700 families in two-story apartment buildings and is served by a central power plant. We offered savings to the housing authority if they would contract with us to operate the plant. We would invest our capital, provide our knowledge, and guarantee savings. The administrator had many responsibilities and could use our help, but since all the savings would simply lower the federal reimbursement each month, there was no gain to the housing authority. However, if we failed in any way, the problem would fall to the local administrator. Small risk, but no reward.

We did finally succeed and took over the plant operation in July 1997. In two weeks, using knowledge and screwdrivers, our team reduced gas use by 25 percent. In four months, we fixed leaks and improved reliability. Now we are building new, much more efficient boilers. Carbon dioxide emissions and energy costs are both down. There are many similar opportunities.

The same is true for the regional administrator of GSA, or for a military base commander, or for the manager of any other government-funded facility. The financial controls over budget are established at a very high level, but the energy-plant decisions are made at a lower level, usually without the capital or expertise to focus properly on efficiency. Either of two approaches is possible to cure this efficiency barrier: (1) mandate efficiency standards for all comparable units or (2) allow the local unit to keep some of the energy savings to fund local programs. Both approaches may be appropriate.

ACTION #1: Initiate executive orders that let housing units and other government-funded installations keep a portion of their savings when they install energy efficiency.

ACTION #2: Executive orders should mandate fossil fuel use standards per square foot or other appropriate unit and allow performance contracting with third parties.

The forty-six-state Department of Defense and Federal Energy Management program initiated in 1997 is pioneering this approach and will increase efficiency. The goal of this program is to obtain private capital to install energy-efficient plants and devices. The DOD has issued authorization to nine preapproved vendor teams to negotiate directly with each military base commander to invest in energy-saving devices, payable out of energy savings. The initial program has a limit on total private investment of $1.1 billion. The bases will be allowed to keep all of the savings. The base facility managers are suddenly very motivated to install energy efficiency.

State Laws Based on Obsolete Technology Prevent Efficiency

Power entrepreneurs are often prevented from installing energy-efficient technology because of state laws that are based on obsolete technology or other unrelated factors. These rules may have been originally passed to ensure public safety or may have been proposed by labor unions to preserve existing jobs. Many of these rules have been made obsolete by new technologies and need to be repealed or severely modified. Here are some examples, drawn from a much longer list.

Example #1: State Law Blocking Efficiency

In 1899, Massachusetts passed a law that required every steam turbine fed by a 1¾ inch or larger steam pipe to have a licensed operator present in the same room, twenty-four hours per day.[109] The technology was in its infancy, and thought to be potentially dangerous. There were no computer controls nor even any computers to shut down the machine in the event of a problem, and there was little experience. Ninety-nine years have passed. Today, mass-produced backpressure turbine generators can be installed in buildings to reduce the pressure of district steam. They convert 97 percent of the energy content of the heat they remove to electricity. After counting the boiler efficiency to make the extra steam, these units have a fossil fuel efficiency of 80 to 82 percent fuel to electricity—2.5 times the national average efficiency of generating electricity. They could save a lot of carbon dioxide and money. A 100-kilowatt turbine that could, without operators, save $20,000 per year and that is allowed to run without operators in forty-nine other states, is required by Massachusetts law to have a full-time operator. Obviously, no efficient, small backpressure turbines are installed in Massachusetts.

In late 1997, when the Massachusetts legislature passed new laws to deregulate electricity, they slightly modified the 1899 law to allow the licensed operator to be out of the room, as long as he or she was on the premises. This small change of law will change nothing else, as the cost of that operator exceeds the savings from installing these efficient devices. There are no remaining safety reasons to mandate operators on small steam turbines. If unattended operation is so dangerous, where is the smoking gun of personal injury in the other forty-nine states? There is no history of safety problems.

Example #2: Making an Example of MIT

When some organization or institution does try to improve its energy efficiency by combining the generation of their heat and power, and they stop purchasing their electricity from the local monopolist, the barriers to efficiency come into play with awesome force. The Massachusetts Institute of Technology recently increased its efficiency by installing a highly efficient 22-megawatt cogeneration plant. However, their ensuing experience put a chill on efficient combined heat and power everywhere. The entire story is reprinted below from the *Northeast Midwest Institute Economic Review*.

> The Massachusetts Institute of Technology (MIT) in 1985 began to consider generating its own electricity for a variety of reasons. With its students now using PCs, stereos, hair dryers, and toaster ovens, the university faced soaring electricity cost from the local utility, Cambridge Electric Company (CelCo). Many of MIT's world class research projects also could be ruined by power quality problems or service interruption. Also, MIT's steam-powered heating and cooling system, which included 1950s-vintage boilers that burned fuel oil, was a major source of local air pollution.
>
> The university selected a 22-megawatt combustion-turbine-based, natural gas-fired, CHP system. The system was to be 18 percent more efficient than generating electricity and steam independently. It was expected to meet 94 percent of MIT's power, heating, and cooling needs, and to cut annual energy bills by 40 percent. MIT expected to recoup its investment in less than seven years.

MIT's first major hurdle was in obtaining the environmental permit needed before construction could begin. Because it retired two 1950s-vintage boilers and relegated the remaining boilers to backup and winter peaking duty, the CHP system would reduce annual pollutant emissions by 45 percent, an amount equal to cutting auto traffic in Cambridge by 13,000 round trips per day. Despite these substantial emissions savings, plant designers had problems meeting the state's NOx (nitrogen oxides—a smog precursor) standard. The state's approved technology for meeting that standard—which was designed for power stations more than ten times larger than MIT's generator—was expensive and posed a potential health risk because of the need to house large amounts of ammonia in the middle of the campus. MIT appealed and won by performing a sophisticated life-cycle assessment that showed its innovative system had lower net emissions than the state-approved technology that vented ammonia.

Although MIT overcame the environmental hurdle and completed construction in September 1995, the story wasn't over. MIT had the misfortune to leave the grid just when Massachusetts was restructuring its electric utility industry. The Massachusetts Department of Public Utilities (DPU) approved CelCo's request for a "customer transition charge" of $3,500 a day ($1.3 million a year) for power MIT would not receive. MIT appealed the ruling in federal court—arguing that it already was paying $1 million per year to CelCo for backup power, and that for ten years CelCo had been informed of the university's plans and could have taken actions to compensate for MIT's self-generation, and that the utility's projected revenue loss was inflated. But the judges ruled they did not have jurisdiction. MIT then appealed to the Massachusetts Supreme Judicial Court, which in September 1997 reversed DPU's approval of the customer transition charge, remanded the case for further proceedings, and stated that no other CelCo ratepayers contemplating self-generation should have to pay similar stranded costs.

In the meantime, the state's proposed restructuring legislation could have raised the amount that MIT would have had to pay for leaving the system to as much as $6.5 million. Fortunately for the university, the Massachusetts law exempted CHP generators from an "exit charge," or non-bypassable transition charge paid by anyone who leaves the system.

Although MIT now has a CHP system—which is saving money and reducing pollution—the university's experience demonstrates the substantial efforts that have been needed to overcome regulatory barriers to efficient combined heat and power plants. MIT was a major research institution ready to fight, but most heat and power users would have had neither the financial resources nor technical expertise to surmount these barriers.[110]

The MIT experience was widely reported and followed by other universities and other would-be cogenerators. MIT finally won its twelve-year-long battle, but the country lost the war for efficiency. These misguided actions sent a chill through the CHP market and have greatly slowed progress toward efficiency.

Example #3: State Law Blocking Efficiency

Many states still have laws requiring licensed operators for any power plant with steam pressures above 15 pounds per square inch, based on the safety technology of the 1930s. New computerized control technology makes these laws obsolete. For the past twenty years, France has allowed power plants without operators, providing they have redundant safety controls. France has not experienced safety problems. These obsolete state safety laws requiring full-time operators discourage combined heat-and-power plants from using small-scale combined-cycle technology, since the steam pressure needed to operate steam turbines forces the owner to add needless operators around the clock. Numbers illustrate the point. If the total cost of a licensed operator, including social security, insurance, vacations, and pension, is $50,000 per year, and the rule forces adding one operator per shift or five persons, the law increases costs by $250,000 per year. Taking account of vacations and sick leave, one must employ five people working forty-hour shifts to cover a facility twenty-four hours per day, seven days per week.

Example #4: State Law Blocking Efficiency

Some states require siting approval by the public service commission for any new power plant, regardless of size. For example, Trigen is currently seeking to build a relatively small 4-megawatt power plant inside its downtown Baltimore steam-production plant, which would dramatically improve efficiency and reduce carbon dioxide. There is, however, a legally required siting review by the Maryland PSC that could take up to eighteen months. It is quite likely that the commission will approve, but a considerable amount of fuel will be needlessly wasted during the eighteen-month review period. This law was clearly intended for regulation of the typical monopoly utility plants that range from 200 to 1000 megawatts but it is applied equally to small on-site plants. If any consumer chooses to install a heat-only boiler plant of any size that is inherently inefficient, no siting review is required. If that same consumer chooses to install an inherently efficient combined heat-and-power plant, many states will require a siting review. How do these rules protect the public? This is yet another barrier to efficiency, based on the central generation paradigm.

Ignoring the cost to society of this delay, it is amusing to see the way the siting reviews have been defended by the monopolists in the past. Trigen is working with Cinergy to develop an efficient combined heat-and-power plant in Ohio. The full output will serve the facility and lower costs. The project requires Ohio power-plant siting review which takes five months. The Cinergy regulatory lawyer has argued for the past ten years (as a monopolist) that all power plants must have siting review. Today, with competition approaching, the same lawyer is now arguing against siting review. This is an example of how even the promise of competition starts to remove old barriers to efficiency.

Example #5: State Law Blocking Efficiency

The states of Louisiana, North Carolina, and Maryland, to name a few, prohibit third parties from owning electric generation and from selling power to others, even to others on the same site. The only legal power plants in these states are those owned by the monopolist, or plants owned by the facility owner. For example, a furniture factory in North Carolina can only: (1) buy power from the monopoly utility or (2) invest in his or her own cogeneration plant. This discourages facility owners from making their operations more energy efficient. Most facility owners are not in the business of power generation and want to invest their capital in their core activities. The furniture plant does not want to tie up $10 million in power production, so it gives up on utilizing cogeneration.

MUCH ENERGY CAN BE SAVED BY RULE CHANGES

The above notes are far from a comprehensive list of the barriers to efficiency; they simply give a flavor of how the paradigm of monopoly electric-only generation has led to inefficiency. Many of the laws blocking efficiency are as obsolete as the Kentucky law, still on the books, that requires all citizens to bathe once a week. Unfortunately for the cost of power and for the global atmosphere, the barriers cited above are enforced. There are enormous efficiencies to be gained by simply eliminating obsolete rules and by formulating regulatory approaches to achieve legitimate societal ends in ways that encourage efficiency.

Include Barriers to Efficiency in Economic Models of Climate-Change Mitigation Costs

This chapter could be useful reading for every macroeconomic modeler who assumes that markets are perfectly efficient—that all economically justified efficiency investments have already been made. Those claiming that climate-change mitigation will wreck the economy cite economic models that purport to show how much reducing carbon dioxide will cost, but every model reviewed by economist Dr. Gary Yohe of Wesleyan University assumes that saving the first ton of carbon dioxide will cost money. My friend Amory Lovins truly captured this thought in a letter to the editor of the December 20, 1997, issue of the *Economist*:

> Sirs: By placing economic theory over practical business experience, you assume an economic problem where none exists. Your logic follows impeccably from a single mistaken premise: if large gains in energy efficiency (the dominant "no regrets" option) were truly profitable, then such measures would have been implemented already. In fact, we have documented scores of egregious market failures that leave some $300 billion of potential annual energy savings in America unrealized.
>
> The empirical reality of large, profitable, energy efficiency inverts your premise. Protecting the climate is not costly but profitable, because saving fuel is generally cheaper than buying it (neglecting any further benefits from not burning it). With market failures corrected, huge energy savings can be quickly bought, even at current price. We

should get the price right, too, but the ability to respond to price is even more vital. Uncertainties about climate are irrelevant: we should buy energy efficiency anyway simply to save money. It does not matter who goes first—but those who do will gain the greatest and earliest rewards. Buying efficiency, far from competing with development needs, frees up vast sums to finance them. The debate shifts from prices and pain to markets, enterprise, innovation, competitive advantage and economic opportunity.

Those theoretical economists who wouldn't pick up a bank note from the street (if it were real, someone would have done so already) will not capture these profits. Alert executives will.

Economists are the usual champions of free markets and deregulation and have provided the intellectual underpinning that justified Congress passing laws to deregulate interstate rail freight, interstate trucking, gas transmission, air travel, and long-distance telephone service. In each case, the quality of service and options offered increased and the cost of service dropped in real terms by 27 to 57 percent.[111] I know of no econometric model that predicted this level of savings in any of the five industries that were deregulated. Yet, when asked to build models to predict the costs of lowering greenhouse gas emissions, the economic model builders seem to ask only the pricing question: How much will the price of fuel have to be increased by new energy taxes to induce behavior changes to lower carbon dioxide emissions?

Their answer frightens most of us, because if we rely on price alone to cause the public to change its behavior so as to conserve fuel and reduce carbon dioxide, we need to dramatically increase energy prices. These economic models assume that all fuel conversion and energy use is optimal at today's energy prices. Their models fail a fundamental test of all good models. They don't explain observed data. The models do not explain how the United Kingdom lowered carbon dioxide emissions from electric production by 39 percent in five years and lowered prices to consumers by 15 to 20 percent. The models don't explain why Trigen and other independent power companies are making money by investing in efficiency, and cutting carbon dioxide in the process. The model builders are answering the wrong question. They seek to find how much prices must increase to mitigate climate change, *holding all else constant.* However, we do not need to hold all else constant. We can remove barriers to efficiency and let the market seek optimal solutions without all of these constraints.

Economists who build models without including obvious ways we can reduce carbon lend credence to the caricature of an economist as an expert who will know tomorrow why the things he/she predicted yesterday did not happen today. A better question for economists to tackle is, How much will society increase energy efficiency and lower carbon dioxide emissions if barriers to efficiency are eliminated?

We have joined others to encourage the Department of Energy's Assistant Secretary for Energy Efficiency and Renewables to find and eliminate these barriers to efficiency wherever possible. I am pleased to report that a strong effort is underway involving the DOE, EPA, DOD, and the Treasury Department. Many barriers could be changed by executive order. Some barriers to efficiency require congressional action. Other barriers can be removed by persuading states and regulatory commissions to opt for efficiency. The carbon dioxide reduction payoff potential for this effort is certainly worthy of the effort. The economy and the environment will reap the rewards.[112]

9

Federal Action to Decarbonize

FINDING ONE NEW SIGNAL THAT AFFECTS ALL ENERGY THINKING

The market would quickly wring out most of today's fuel conversion and energy use inefficiencies if government would immediately: (1) deregulate electricity everywhere; (2) eliminate the barriers to efficiency embodied in tax, environmental, and regulatory law; and (3) eliminate the present subsidies for fossil fuel. These rules cause our society either to burn more fossil fuel than is necessary to produce the required energy products, or to remove the incentives for energy professionals to help consumers to stop wasting energy. These rules keep society from earning a good payback by investing in knowledge and capital to replace fossil fuel use. These rules weaken our economy; eliminating the rules would strengthen the economy and lead to rapid decarbonization. Unfortunately, it is not possible for the federal government to pass one law to accomplish these three goals.

There are many reasons why clean and comprehensive deregulation is politically unlikely. Fifty states have monopoly protection laws covering electric generation, distribution, and sale. Note that ten states adopted some form of deregulation law during 1997 that will end parts of the present monopoly protection, in time, but none of the laws are clean ends to generation monopoly protection, and all keep distribution monopolies intact.

These monopoly protection laws were not in the past deemed to violate the commerce clause of the United States Constitution, but technology and conditions have changed. I think a current analysis would overturn the monopoly protection laws. The Supreme Court has long recognized the right of states to use their police powers to regulate business "affected with a public interest." But this has presumed that electric generation and distribution are strictly *intrastate* activities. The reality is that these activities today directly affect *interstate* commerce and should be subject to preemptive federal jurisdiction under the Commerce Clause.

The state laws have been allowed to stand because society wanted to electrify rapidly. By now these laws that restrain trade between states are ninety years old. Overriding these laws with new federal legislation invokes arguments about states' rights, and this makes comprehensive national deregulation of electricity difficult. The existing municipal electric monopolies, Rural Electrification Utilities, and Federal Power Marketing Agencies are vested interests that fear the impact of competition and resist change. Roughly 25 percent of United States electric consumers receive government-subsidized electricity and are naturally loath to lose their subsidies. Cities and states use local monopoly utilities as tax collectors. If federal legislation ends monopoly protection, these various governments will lose tax revenue and face the painful process of raising other revenue. Then the providers of heat and power would be taxed like all other businesses, which is usually at a lower rate than applied to utilities. The higher than normal taxes on monopolies generate significant revenues and create an obstacle to removing monopoly protection.

Likewise, eliminating all of the barriers to efficiency with one piece of federal legislation is virtually impossible. The legislative process uses separate congressional committees to work through the details in different areas, and important barriers to efficiency are found in environmental law, tax codes, securities and exchange rules, procurement policies, and other areas. Each relevant congressional committee would be required to examine its own area and determine how to end the barriers to efficiency embodied in the laws and regulations it supervises. Each committee would need to be sure that after the regulations were cleansed of barriers to efficiency, that they still accomplished their original purposes.

The executive branch presents the same problem. Separate cabinet departments have responsibility for housing, military bases, federal facilities, environmental protection, etc, and each has built up rules and approaches that unthinkingly assume a central generation paradigm. The regulations of each of these departments typically fail to distinguish between energy use and energy-conversion efficiency. Each cabinet department must have its specialists examine all of their programs to surgically remove the barriers to efficiency while preserving the intended results of their programs.

Executive-branch cabinet departments and the legislative branch are unlikely to spend the time and effort to clean up all of the barriers to efficiency without public and private pressure. They have many other priorities, and they face another obstacle. These agencies and committees are all, to some degree, captives of their constituencies, and will not eliminate the barriers to efficiency unless there is pressure from those constituencies. That pressure won't exist until everyone understands how inefficient we are and what that inefficiency costs the economy and the environment.

The third area of change is no easier. Eliminating the subsidies for fossil fuel is also politically difficult. Oil-patch senators and representatives have fought hard for the oil and gas depletion allowances that subsidize petroleum businesses in their states. As an example of how difficult it is to end subsidies, consider the tobacco farmers. The Clinton administration has aggressively attacked cigarette smoking as damaging to people's health and damaging to the economy due to the medical costs of smoking-related illness. However, ending subsidies to tobacco farmers is another matter. Ending subsidies on fossil energy production will be just as tough.

Separate actions can and should be taken to whittle down the number of laws, regulations, and rules that encourage profligate use of fossil fuel. It is sensible to phase out the estimated $5 to $25 billion of fossil fuel subsidies per year, given mostly to the oil and gas industry.[113] It makes sense to address separately each of the barriers to efficiency identified in Chapter 8. It makes sense to educate and to create incentives for consumers to make more efficient use of energy. The Clinton administration budget for fiscal year 1999 makes some of those proposals. However, achieving changes to so many laws is unlikely until pressure builds to stop blocking efficiency. Policy makers are searching for a national approach to inducing efficiency that is possible to enact.

FORCES BLOCKING DEREGULATION OF ELECTRICITY

National deregulation of all electricity would unleash competition and provide an understandable and predictable set of rules for all energy companies. Giving all consumers freedom of choice to pick an electric supplier would allow market forces to go to work and destroy inefficiencies in today's system, but we probably cannot get there from here. There are strong political forces at work that will not yield to such a frontal approach, no matter how valid.

States' Rights Block National Action

States' rights are a deep and enduring part of U.S. federalism; every attempt by the federal government to usurp powers from the states results in spirited political debate. This is true of proposed federal laws to deregulate electricity even though the original granting of monopoly power to the electrical industry no longer has a legitimate basis for support. Every aspect of the electric industry of the 1990s either is directly in interstate commerce or directly affects interstate commerce. Nonetheless, each state has been allowed to regulate electricity for the past seventy to ninety years, and this is now considered a sacred states' right, granted in the Tenth Amendment to the United States Constitution. Each state gains certain advantages by having its own public service commission, having its politicians supported by the regulated utility, and having the power over electricity. Peter Bradford, who has served as the chairman of both the Maine and New York State regulatory commissions and is now a consultant on regulatory matters, recently told me of his belief that deregulation is intimately linked to campaign-finance reform. On questioning, Peter indicated that in conversation after conversation with state legislators, he hears about the vise that these elected officials are caught in if they oppose the incumbent monopoly and risk losing utility campaign contributions to their next election bid. The states will not open electric markets fully without a big fight that will take years. Meanwhile, society will go on investing in inefficiency and continue polluting the air.

Rural Electrification Act

As a part of the drive to electrify the nation, the federal government created the Rural Electrification Administration (now the Rural Utilities Service,
or RUS) in 1936 to electrify rural America with subsidized loans and grants
to rural electric cooperatives. At the time of the act, only a fraction of farms
and rural households had access to electricity. The act enabled REAs around
the country to use federally guaranteed money to bring power to remote
farmers. Currently, the highest rate paid by an electric co-op borrower is
the lesser of 7 percent or the rate the United States Treasury pays to
borrow.[114] This is a significant subsidy to one-time rural areas by the American taxpayer. Those enjoying subsidized power vote, and as farmers enjoy
some special political leverage. They can be counted on to oppose any
market opening that is seen as a threat by their local REA or RUS.

Municipal Power Companies

There are 2,014 cities that have municipal power companies that distribute power to the people in their city. These municipal power authorities are owned by the cities they serve, and are typically governed by
boards that are appointed by the mayors or city councils. Board seats,
and sometimes managers' jobs, can be patronage jobs—a way to reward
folks who have supported the local politicians in prior elections. Roughly
625 of these municipal electric companies also generate some or all of
the electricity they need, while the rest purchase power from investor-
owned utilities or from federal power authorities. They all employ people
who vote. The people so employed fear competition (just like all of us)
and are fighting hard to keep their monopoly positions.

   By the terms of their charters, municipal utilities have no way to sell
services in other territories, so deregulation is seen as only a threat, with
no possibility of gain. (Investor-owned utilities have upside potential from
deregulation as well as threat because they can expand elsewhere.) The
municipals and their associations are at work on the national political
scene to try to block any legislation that allows their customers to have
a free choice of electric supplier. In California, retail customers of
investor-owned utilities gained the freedom to choose their electric supplier on March 31, 1998, but the customers of municipal utilities were
denied the choice until four years later.

Federal Power Agencies

Another big obstacle to deregulation is public power and the many vested interests who gain by preserving the monopoly-protected status quo. Charles Bayless, the chairman, president, and CEO of Illinova Power, recently wrote a paper titled "Public Power—Time's Up," in which he gave this description:

> In the years following the Great Depression, the demand for job creation and rural electrification dictated a time to build public power. Agencies such as the Tennessee Valley Authority and countless rural cooperatives and municipal utilities were formed or expanded and performed extraordinary tasks in providing jobs and bringing electric power to rural regions. Today, most of the agencies formed during the Great Depression, such as the WPA and the CCC, have long passed into oblivion, having fulfilled their purpose. Given the recent movement to competition in the electric power industry, the same should be true for some segments of public power.
>
> Three factors indicate the time for public ownership in the generation segment of the power market has passed. First, it is unfair for federally subsidized agencies to compete in the competitive market for electric generation. Second, there is no remaining public policy need which public power fills—the public power mission has been largely fulfilled. Third, given the growing concern over government budget deficits, it is no longer necessary or feasible for federal, state and local governments to provide a subsidy of more than $11 billion per year. In this respect, public power agencies have become one of the largest wealth transfer schemes in American history, costing all taxpayers billions of dollars but providing benefits to less than 25 percent of all Americans.[115]

As Bayless says, "Time's up."

Look at the above description and think about how the existence of public power will impact a congressional debate over deregulating the entire U.S. electric industry. The 25 percent of Americans who receive these nice subsidies from the remaining 75 percent of Americans have clear economic interests in maintaining the status quo. In politics, the minority groups who stand to gain or lose rally political support for their causes. The majority group, who pays the tab, cannot effectively mobilize to fight each of these organized minority groups, and this makes it very hard to change old and bad laws. Some wit referred to this phenomenon as the "tyranny of the minority."

The difficulty of ending obsolete public power subsidies involves both throwing out the old central generation paradigm and overcoming a self-interested minority. To change our wasteful energy system, we must overcome the undoubtedly strong efforts of the subsidized public power groups. A competitive market will, in time, deliver the customers of public power cheaper electricity than they purchase today from their subsidized suppliers, because those suppliers are very inefficient and wasteful, but the public power customers do not know this.

Utilities as Tax Collectors

The final factor making national deregulation difficult is the way states and cities have used the monopoly utilities as tax collectors. A practice as old as government is to find a commodity that all people must have, then subject that commodity to government control and tax its use. Romans throughout their empire, and later the British colonial government in India collected significant government revenues by taxing salt. King George tried to pay for his colonial adventures with a tax on tea, which resulted in some early pollution of the Boston Harbor when irate colonists staged the Boston Tea Party. A national program to open all electric sales to competition has no way of dealing with each state's unique approach to raising revenues by taxing electric power production. Besides sales taxes on power, many states impose a gross-receipts tax on monopoly sales, requiring the monopoly to pay a percentage of all revenues they collect as tax. Other states have special property taxes on utility-owned generation and transmission assets that are significantly higher than property taxes on other types of real estate. In a deregulated market, many institutional buildings and companies will generate their own power and this will lower state-tax revenues. Competition will always seek ways around taxation, and monopoly protection keeps electricity an easily taxable commodity.[116]

An example of the local difficulty is Trigen Energy Corporation's Baltimore district steam system, which under Maryland law is a regulated utility. In 1998 the state legislature considered a bill to deregulate the steam system and this raised a taxation concern. We pay roughly $300,000 more per year in taxes to the city of Baltimore than we would under the property-tax rules that apply to non-utilities. The assumption, which was always false in the case of a steam system, was that the system was a monopoly, and since the customers had no other choice, added taxes would not cause them to leave the system. In fact, the customers have other choices, like installing their own boiler and burning gas or oil to make their thermal energy, and some have exercised this choice. We worked around this issue with the mayor of Baltimore, but on the final day of the 1998 session, the monopoly-utility lobbyist managed to have the bill amended to deregulate electricity as well, and the bill was defeated.

To summarize, ninety years of monopoly protection and a long history of government actions to electrify society have left a patchwork of special interests that can be counted on to oppose full electric deregulation. Perhaps, as the public debate goes forward, people will learn more about the damage to the economy and the environment by the state's exercise of its police power to prevent competition in electricity. Perhaps people will become more aware of how we make less than optimal energy decisions without competition. But, as our then three-year-old son used to say in response to my wife's loose promises, "Maybe not." Meantime, the carbon dioxide concentration in the atmosphere grows, and every day we waste money and fuel, locking ourselves into investments in power and heat plants that do not make economic or environmental sense.

HOW TO CHANGE POWER-PRODUCER BEHAVIOR?

There has been some thought given to ideas for a national action that could overcome some of these barriers to efficiency and cause power companies' behavior to change. Before we examine the three current approaches to induce decarbonization and suggest a fourth action, it is essential to look at the behavior that must be changed.

As noted earlier, there are three ways to decarbonize the economy. Two ways will maintain or increase everyone's standard of living. The first way to decarbonize is to change consumer use of energy to lessen waste. The second way to decarbonize is to change the decisions and actions of the fuel-conversion industry so it produces heat and power with less fossil fuel, either by increasing efficiency or by relying on more renewable energy. The third way to decarbonize is feared by many—either lower the standard of living or reduce global population. Are there one or more national actions that would lead society toward both of the first two ways of decarbonizing?

The first possible way to decarbonize and maintain the standard of living involves consumer behavior and the second involves changing the behavior of the energy-conversion industry. Most actions do not effectively address both behaviors, because the motivations and capacity to modify behavior are different for the two target groups. Actions that will educate and motivate consumers to save energy are very different from actions that will motivate the energy-conversion industry to be more efficient and use less fossil fuel.

If deregulating all electric markets cleanly is improbable, what are the alternatives? The challenge is to find elegantly simple federal actions that mobilize the market.

Three national actions that are currently being discussed are mistakenly assumed to impact both consumers and fuel converters, ignoring the difference in motivations of the two sets of decision makers. With this difference in motivation in mind, we will look at the three national actions currently proposed to engender decarbonization and suggest a fourth possibility—a national Fossil Fuel Efficiency Standard that would encourage fuel converters to decarbonize.

THREE CURRENT FEDERAL PROPOSALS TO DECARBONIZE

In the debate about climate-change mitigation and a sustainable future, three approaches to decarbonize the economy are being discussed by the federal and/or state governments. The negotiators involved in the United Nations Climate Change negotiations like to talk of a "carbon cap." By this they mean that each nation would agree to cap its carbon emissions at some agreed level. The Kyoto Protocol language says that industrialized nations will cap their carbon emissions by the year 2010 at levels 5.38 percent below the 1990 carbon-emissions level.

A second proposed federal action, recently supported by the Alliance to Save Energy,[117] is to apply revenue-neutral taxes to energy. This action would increase the tax on fossil fuel—thus increasing consumer incentives to reduce fuel use—but would lower other taxes so that total federal tax revenue would remain constant.

Some state deregulation legislation, several proposed federal deregulation bills, and the Clinton administration's Comprehensive Electricity Competition Plan of March 25, 1998, all require a "renewable portfolio standard." This would require each seller of electricity to include a specified set percentage of power generated with nonfossil fuels. These renewable portfolio proposals would give a special status to renewable energy.

Carbon Cap—Impact and Problems

Lowering the net carbon emissions by emitting less carbon and/or sequestering more carbon by planting trees or taking other actions is the desired result of any proposal to mitigate climate change. Regardless of the mechanism used to change behavior of all sectors of the economy, the goal must be to lower the carbon dioxide emissions from burning fossil fuel. There is a comforting ring about a cap on carbon to the senior policy advisors who negotiated the Kyoto Protocol, because it seems to encompass what is needed. But passing a law that puts a cap on carbon emissions tells us nothing about how to change behavior to lower fossil carbon use. Any carbon-cap law must consider how to allocate the carbon emissions across the economy, how to measure and enforce carbon dioxide reduction, and how to change the behavior of consumers and fuel-conversion companies.

The carbon-cap proponents envisage a system of allocation of carbon emissions to each power plant or boiler. Each power plant would be required to limit emissions to their carbon dioxide limit, or to trade for credits with other plants that emit less than their limit. This logic follows the success of the trading of sulfur emission caps that was instituted in 1992. The market found the cheapest ways to reduce sulfur, and this dropped the cost of removing sulfur in power-plant exhaust from early estimates of $1,200 to $1,500 per ton to $100 per ton today. The market worked to lower the cost of removing sulfur, but there were terrible inequities in the allocation of rights to emit. Sulfur emissions were allocated on the basis of three prior years of emissions, so that the worst polluters received the largest allocations, the relatively clean power plants received the lowest allocations, and those proposing new plants that had a better way to make power or heat received no credits at all.

The innovators with no history have to buy rights to pollute from a historical polluter. If the carbon cap is allocated based on historical emissions, it will repeat this mistake.

The fuel-conversion industry does not create demand for energy, it merely supplies the energy demanded by consumers. Consequently, to be fair, a carbon cap must be allocated equally to all energy produced. If this happens, energy converters will be free to choose technologies and fuels that produce heat and power with the allowable carbon. An output-based allocation will let the market work, and will reduce CO_2. If energy demand rises, the allocation per unit of energy produced can be further lowered.

A further issue with a carbon cap is that carbon emissions come from millions of sources, many of which are neither regulated nor monitored. A comprehensive carbon cap is simply not feasible, as it would require that every boiler have a way to measure fuel used and report carbon emissions based on the fuel burned. What do the owners of small commercial boilers do to reduce carbon from their boilers? It is prohibitively expensive to monitor every home furnace or boiler. While all electric production and most industrial and institutional thermal production is from power plants that are monitored and regulated by the local environmental-protection officials, the carbon emissions from home and small business boilers are not monitored. The carbon emissions from automobiles, which represent one-third of the United States total, are not only unmonitored, they are virtually unchangeable, once the car leaves the factory. No law will regulate all carbon emissions.

A final issue with the carbon-cap approach is the need to separately address the behavior of two relevant sectors—the fuel conversion industry, and consumers. The stimulus and actions that will cause the fuel-conversion industry to adopt more fossil efficient practices are not the same stimulus and actions that will cause consumers of heat and power to be more efficient in their use. Chapter 7 suggests some potential actions to change consumers' attitudes toward energy and to encourage consumers to use more efficient appliances, houses, automobiles, and offices. These educational actions and tax credits for efficient appliances will change consumer behavior but will have minimal impact on the way the fuel-conversion industry operates. On the other hand, a carbon cap, which could force fuel-conversion industry changes, will, if applied to all consumers, be unimaginably cumbersome to administer and enforce.

A carbon cap applied to large-scale fuel converters is complex, but conceivable. There are relatively few power plants and they must all report their fuel use and output to the Energy Information Agency. The large industrial boilers and industrial cogeneration plants are all subject to compliance with Clean Air Act standards, and also monitor and report some aspects of their operation. The problem is one of motivation. If a monopoly utility is forced to reduce CO_2, it could improve efficiency. It could not, however, recover its investment in efficiency without cumbersome rate hearings. So why not remain inefficient, buy permits to emit carbon, and pass the cost on to the utility customer? With no hearing required, and no incentives to invest in efficiencies, the utilities will keep their efficiency closer to 33 percent than the 80 percent that is economically feasible.

Revenue-Neutral Taxes on Fossil Fuel or Energy

A second option being considered by policy makers to encourage decarbonization of the economy is the application of new taxes on fossil fuel, or on energy, to increase the price of energy and cause people to invest more money in efficiency. To keep from disrupting the economy, proponents call for offsetting reductions of other taxes so the energy taxes are revenue-neutral. The logic for higher energy taxes is consistent with Economics 101. Economists observe that consumers can make substitutions for any good or service. When the price of any item or service increases, consumers' other choices become more attractive. For example, a tax on oil and gas would make it more expensive to heat your home, so you might consider investing in added insulation or in controls that will create zones in your house and allow each zone to be heated only when occupied. As the fuel tax goes higher, the incentive to invest in efficiency grows.

The Alliance to Save Energy, a not-for-profit organization, supported by many corporations and individuals, issued a study of the cheapest way to decarbonize the economy in January 1988. *Price It Right* argues that revenue-neutral taxes on energy that cover the full social costs of burning fuel were a less costly way to induce behavior changes than mandated carbon caps with "command-and-control" enforcement structures.[118] This finding is almost certainly true, but is it relevant? Both approaches have costs while deregulation will produce savings.

The study did not differentiate between consumers of energy and the fuel-conversion industry. This is a fatal flaw. Added taxes on fuel raise the cost of fuel purchased by a monopoly utility, but the regulators allow the cost of fuel to be passed through to the customers. Purchasers of electricity would pay higher prices, and this might impact their decisions to purchase more efficient appliances and air conditioners. However, the Alliance to Save Energy study also assumes that taxes would cause regulated utilities to change their behavior, and this seems unlikely. As noted earlier, the very strong increase in oil and gas prices that resulted from OPEC's two major shocks did not increase utility efficiency above its dismal 33 percent.

We could expect monopolists to pay lip service to becoming more efficient, and to make some token investments, but we must remember how they are regulated. As with investments to meet a carbon cap, the regulated utility will have difficulty, under current regulatory policy, recovering the investments or earning a profit from those investments. Chances are, they will simply pay the tax and pass the added cost on to electric customers. If the taxes are high enough, some of the customers will make investments in efficiency and purchase less electricity, but this long-term impact won't suddenly lower the utility's profits. Monopolists are not stupid. Who would invest in efficiency with no guarantee of a return? They will pay the tax.

Higher energy taxes will have an important impact on industrial and commercial power plants that produce thermal energy for processes, institutions, and offices. Since these sectors face competitive pressures and as a result have no way to automatically pass added energy costs through to customers, they will respond to energy taxes by substituting capital and knowledge for fuel. Industrial firms constantly look for efficiency investments and calculate payback and return on their investment. A tax on energy will make the energy saved more valuable and hence make the payback periods on energy-efficiency investments shorter, the return on investment higher, and will cause added efficiency. This aspect of the tax idea has merit in the industrial sector, but further complications block the largest potential efficiency gains. The most promising way to increase the efficiency of producing thermal energy for commercial, industrial, and institutional users is to combine the generation of heat and power—to capture the normally wasted heat from electric generation and use it for the needed thermal energy. But monopoly rules forbid selling electricity to others and in some states forbid third parties from even generating electricity. This powerful efficiency technique of combining heat-and-power generation will not respond very strongly to added taxes until many barriers are removed.

Renewable Energy Portfolio

Representative Dan Schaefer from Colorado heads the United States House Subcommittee on Energy and Power, a part of the House Commerce Committee. He introduced H.R.655, "The Electric Consumers' Power to Choose Act" in 1997 to federally deregulate electric production. The bill establishes a renewable energy requirement, expanding the federal role in determining the mix of electricity generation. Congressman Schaefer's bill requires that by the year 2010, 4 percent of electricity from each seller would come from renewable sources.

The Clinton Administration Comprehensive Electricity Competition Plan calls for 5.5 percent from renewable sources by 2010. The Federal Energy Regulatory Commission (FERC) would be given the responsibility of administering a renewable energy credit program and certifying performance of each generator. These renewable portfolio standards, if passed, would decarbonize the economy to some extent by forcing a portion of all electricity sold to come from renewables. Proponents of this approach hope that the increased demand for renewable energy will move renewable technologies up the learning curve. This would make renewable energy more economical and more able to compete with fossil energy. Renewable energy will become cheaper if either bill becomes law, because renewable energy companies will compete with each other to gain market share, and will invest in both technology development and in mass production.

The various renewable energy portfolio standards give remarkable proof to our contention that we are stuck in an inefficient monopoly-electric-generation paradigm. Ask why we should demand that some electricity must come from renewables, and the answer is that this will lower carbon dioxide emissions. Ask, why not let the market choose whether to burn fossil fuel and use controls to reduce carbon dioxide or rely on renewable energy sources, and the answer is that there are no controls that will efficiently remove carbon dioxide once the fossil fuel is burned. Ask, why not encourage the more efficient use of fossil fuel as an alternative to relying on renewables, and the usual answer is a blank stare.

Policy makers have started with an unexamined assumption that current power generation has reached its peak, and the only way left to decarbonize is to force people to switch to renewable energy. If policy makers start with an understanding that the efficiency of power generation could be doubled while saving money, the answer is different. The renewable energy portfolio standard is clearly a subsidy to renewable energy forms, and does not force those decarbonization technologies to compete with other strategies to decarbonize. By ignoring efficiency and not allowing it to compete with renewable energy as a way to decarbonize, a renewable energy portfolio will cause the prices for power to be higher than would be the case if the market could choose the best way to decarbonize.

ECONOMY VERSUS ENVIRONMENT—FINDING A MIDDLE GROUND

Like any great contentious debate, the proponents of the cap-and-trade approach, the renewables-portfolio-standard camp, the free-market camp, and the "don't-rock-the-boat" camp are organized and vocal. Each camp believes its approach is the best and the mountains of reports, flyers, advertisements, e-mails, and videos issued are rivaled only by the volumes of materials distributed by the other camps. I truly believe that an absolutely unencumbered market for energy would go a long way to increasing the efficiency of producing heat and power and will lower energy waste and energy cost. However, unfettered markets will leave us with a "tragedy-of-the-commons" issue. In free markets without any government guidance, power entrepreneurs will choose the approach that produces the lowest-cost kilowatt-hour. Government's legitimate role is to guide this process toward less pollution and toward a sustainable future.

What is essential is to find the middle ground—the consensus that benefits from differing views. We must find a middle ground between those who understand and appreciate the power of markets and those who understand and appreciate the value and necessity of a healthy environment. We need a middle ground between those who appreciate the many benefits of a growing economy and rising economic standard of living and those who are willing to make deep economic sacrifices to protect the environment. We must find a middle ground between those who favor instantly repealing every obsolete law and regulation and those who want time to adjust to new conditions. We must find ways to blend these views, taking the best from each.

We cannot allow either side of these debates to let their ideal become the enemy of the good. Truly open energy markets and the elimination of all the existing barriers to efficiency will take decades to develop. Something must be done to jumpstart the process of moving our energy sector toward efficiency and sustainability that avoids the political logjam of fighting the coal and petroleum interests head-on, avoids "command-and-control" approaches, provides incentives for the continued commercialization of renewable energy technologies, and allows the market forces to efficiently make the needed changes. Find this approach and the environment will improve as the economy grows.

A FOSSIL FUEL EFFICIENCY STANDARD

A single change to federal law regulating producers of electricity could powerfully eradicate many barriers to efficiency and double the efficiency of U.S. power generation over twenty years. This single action is a mandated Fossil Fuel Efficiency Standard that would apply to every generator of electricity in the nation. It would set a nationwide allowance for the average fossil fuel used per unit of energy produced and would also include a schedule of cutting fossil allowances in half over the next twenty years. Each year, this nationwide allowance for average fossil fuel used per unit of energy produced would be less than the year before.

This proposal is a variant of what is left of a carbon-cap approach, after all the practical difficulties of measuring and regulating carbon from home and commercial boilers and furnaces and cars and trucks are recognized as impossible. Treat consumers with separate educational and incentive programs, but guide the power-conversion industry away from fossil fuel waste.

The formulation is as follows. Measure all input fossil fuel in megawatt-hours of energy content, which is already done and is the basis for fuel sales, and measure all useful output of electric power and heat from every generating plant in megawatt-hours. The national average for 1996 was 2.07-megawatt-hours of fossil fuel energy consumed for every megawatt-hour of heat and power produced. Any increase in efficiency that results in less fuel or more power output from the same amount of

fuel will improve a plant's fossil fuel efficiency. Any production of electricity with wind, solar, nuclear energy, or hydro would increase the
megawatt-hours of output, but would not increase the megawatt-hours
of fossil fuel input, so would improve the fossil fuel efficiency. Recovering
heat from an existing power plant and then selling that heat to nearby
industry for process steam or to nearby buildings for heat or to drive
chilling machines would increase the megawatt-hours of output without
increasing the megawatt-hours of input, so would increase the fossil fuel
efficiency.

Power producers would have to meet the standard for the average
of their total electric production. They could comply in one of seven
ways, including:

- Increase the efficiency of existing generation.

- Recover and sell presently wasted heat.

- Add some renewable energy generation.

- Add new fossil generation that is more efficient than the standard.

- Purchase notional credits for generation from other more fossil-
  efficient producers.

- Close down the least efficient fossil fuel-based generation.

- Pay fines for noncompliance with the standard.

Such a law focuses on the motivation and incentives of fuel converters, but gives them ample time to adopt new energy strategies that
they would adopt anyway but for the barriers to efficiency.

Figure 16 repeats the data on historical U.S. fossil fuel use per
megawatt-hour of electricity, then projects the next twenty years of
fossil fuel per megawatt-hour assuming enactment of the Fossil Fuel Efficiency Standard. We propose a standard that declines on schedule over
twenty years to one megawatt-hour of fossil fuel per megawatt-hour of
electricity produced. The graph shows that this standard would simply
continue and slightly accelerate the trend of the last forty-eight years.

It is possible with today's technology to double the delivered efficiency of all fossil-fuel-produced electricity to 66 percent. A national
average of one megawatt-hour of fossil fuel per megawatt-hour of elec-

Fig. 16. Historical and Projected U.S. Fossil Fuel Use
Per Megawatt-Hour of Electricity

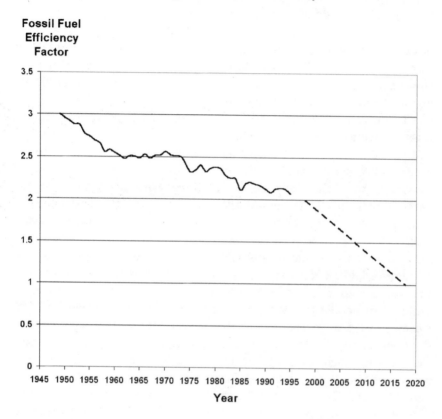

tricity produced could be achieved in twenty years by various combinations of fossil efficiency and more renewable and nuclear energy.

Assume that government's efforts to educate consumers, to conserve energy, and the programs proposed in chapter 7 have generated enough negawatts to keep overall electric load growth minimal. Then the task of improving the nation's stock of power plants will be primarily one of replacing old plants that are alive today due only to monopoly protection.

CAN A FOSSIL FUEL EFFICIENCY STANDARD REDUCE CARBON DIOXIDE AND SAVE MONEY?

These thoughts have filled my brain for more than a year, ever since former Secretary of Energy Federico Peña challenged the presidents of thirty companies to identify actions that would mitigate climate change but would not disrupt the economy.[119]

Trigen's entire business is investing in efficiency, and we have shown that it is economically feasible to produce heat and power with less fuel. We have proven that efficiency pays—if one is persistent enough to get around all of the hurdles and dodges enough bullets. Progress has been slow because of the many barriers, and many pundits seem to accept without question that increasing efficiency to reduce carbon dioxide and save fuel will raise the price of energy and lower our standard of living. There is, however, a growing belief among energy and environmental policy advisors that efficiency is held back by barriers—barriers that can be removed. Mark Hall and I were invited in September 1997 to explain the barriers to efficiency that Trigen has encountered to a group of senior staff people in the DOE's Department of Renewables and Efficiency, headed by Joseph Romm, Acting Deputy Assistant Secretary for Renewables and Efficiency. This session blossomed into a paper ("Barriers to Fuel Efficiency") that formed the basis for chapter 8 herein. The paper was widely circulated and formed part of the rationale for President Clinton's October 22, 1998, policy speech on climate change and has spurred an ongoing DOE effort to identify and remove barriers to efficiency. It was selected by Common Purpose as the best policy suggestion of 1997, indicating some resonance to the idea that efficiency is the issue and that obsolete policies and regulations are real barriers.

Many government officials and policy advisors find the examples of doubling energy-conversion efficiency interesting, but assume that since the major players in the energy-conversion industry have not embraced such efficiency, then doubling efficiency nationally is impossible. They are underestimating the effect of present barriers as well as the power of open markets to find efficient solutions.

It is the mistaken notion of virtually all of the regulatory commissions that they serve consumers' interests by forcing utilities to pass on

100 percent of all energy savings to those consumers. The result is no energy savings. Their focus is on limiting utility profits. The conventional wisdom that electricity is a natural monopoly is wrong, and most policy advisors now know this. However, they have used the monopoly regulation to carry out other goals like cleaning the environment and do not appreciate that the market could accomplish these tasks more efficiently if guided. Some environmentalists assume electric competition will destroy the environment because everyone will buy power from dirty old coal plants. They underestimate market power. Others fear competition will lead to unreliable electric supplies and to poor people being charged much higher electric prices. This has not happened in Great Britain after seven years of market competition, but people argue that the United Kingdom was a special case, motivated by a desire to close expensive coal mines.

All of these pieces of the paradigm are, in our experience, clearly wrong, yet they are all firmly in place. Change is coming slowly with state-by-state deregulation plans, each more convoluted than the last. The deregulation laws enacted to date bend over backward to keep all of the vested interests whole. The vested interests craft eloquent arguments for preserving their favored status. This makes it hard for Congress to throw out the failed paradigm and find the truth.

The Fossil Fuel Efficiency Standard is, in fact, a form of a carbon cap, and is subject to the same issues. Its uniqueness is its focus on efficiency and its refusal to select among fossil fuels. Carbon cap sounds like invasive government regulation. Many people won't perceive any relationship between caps and efficiency.

Invert the carbon cap, convert it to an efficiency standard, and the political debate changes. It is much harder for vested interests to argue against efficiency, especially when the proposed Fossil Fuel Efficiency Standard applies to every generator equally. An efficiency standard alone does not open all electric sectors to competition, but it can force every municipal, REA, federal power agency, IPP, and monopoly utility to adopt the efficiency strategies they would choose under competition. It will force each generator to confront the barriers to efficiency, and they will each start pressuring their regulators to remove those barriers.

This federal Fossil Fuel Efficiency Standard suggestion has emerged over the past nine months and has received countless inputs from a diverse group of reviewers. A Brookings Institute tax expert looked at an early draft and suggested problems and concerns. Tim Wirth, then Deputy Undersecretary of State for global warming, gave the proposal to the International Institute of Education whose energy group under Director Ahmad Ghamarian sought comments from ex-FERC commissioner Charles Stalone, Milton Klein, Joseph Falcone, and others. Mark Hall, Trigen's Director of Government Affairs, has discussed the related issues with many government officials, industry representatives, and environmentalists and has drawn the fine distinctions between efficiency standard and carbon caps. It is no longer possible to affix credit to the various parts of the idea, but what has emerged has become simple in concept and execution.

Charles Bayless and I crafted an Op-Ed piece reflecting these many discussions on the Fossil Fuel Efficiency Standard idea, which was printed in the *Washington Post* on December 31, 1997.[120]

> As world representatives in Kyoto searched for agreement on how to roll back global warming and head off climate catastrophe, they confronted heavy weather themselves. The turbulent discussions put one in mind of Snoopy's observation that "It was a dark and stormy night. . . ."
>
> Our companies burn fossil and other fuels to create electricity, heating and cooling. There is no question in our minds that global warming is a potentially devastating problem and that humankind must reduce emissions of carbon dioxide. As power producers, we understand how much unnecessary carbon dioxide this nation emits, and we know how to cut fuel waste in heat and power generation.
>
> Negotiators from the developed countries, including the United States, went to the table seeking absolute caps on carbon dioxide emissions for each country. This approach drew fire from most of the developing countries because they refuse to cap their future economic growth. Many at home object to caps unless all countries agree, fearing that the United States will lose its competitive position. A treaty was finally negotiated, but lacks support of China and India. To allocate carbon dioxide emissions among power producers and other carbon dioxide emitters will be cumbersome and arduous, and raises nearly insurmountable fairness problems.
>
> As we struggle to obtain final agreement abroad, and work to understand how to implement the accord at home, we should seriously consider a simple mechanism to allocate emissions: Set a standard of fossil fuel use for every unit of heat and power produced. Tighten the standard every year. This standard will lead the power industry to

deploy more energy-efficient heat and power plants and to develop renewable energy from solar, hydro, wind and bio-mass.

Fossil fuel used per unit of heat and per unit of electricity produced is easy to measure and lends itself to simple enforcement. This standard answers the difficult question of who will have rights to emit. Any producer of heat and/or power would automatically receive credit equal to the standard. Those not meeting the standard in a given year would have to follow one of the seven strategies listed above. There is, however, one problem.

Seventy years of monopoly "protection" have severely impeded innovations in the United States electric utility industry. An average American power plant burns three units of fuel to produce only one unit of electricity, venting the other two as heat. In effect, two-thirds of every coal mine is a wasted hole in the ground. It's as much wasted energy as Japan uses each year to fuel its entire economy. And it's a huge amount of money wasted on fuel. Consumers pay for the wasted fuel and the atmosphere bears the pollution.

Anticipating the competitive pressures of electricity deregulation, an increasing number of utilities are adopting new strategies. For example, our joint venture in Golden, Colorado, serving Coors brewery, achieves more than double the average U.S. efficiency. We convert 70 percent of the fuel to useful energy—electricity, steam and chilled water. We cut carbon dioxide in half, save enough fuel to cut Coor's costs and make money. Our entire industry would do better but for outmoded regulations.

Eliminate monopoly rules and entrepreneurs will revolutionize the power industry. Guide the emerging competition with a fossil fuel per unit of heat and power standard that tightens according to a predetermined schedule. Entrepreneurs will then replace or retrofit our obsolete electric-only generation to capture waste heat and produce valuable products such as steam, hot and chilled water. Entrepreneurs will develop and deploy renewable energy. These market forces will lower U.S. costs of heat and power. Developing nations that don't adopt similar efficiency standards will fall further behind the increasingly efficient U.S.

The Kyoto Protocol commits the United States to reduce greenhouse gas emissions by 7 percent below 1990 levels. This is achievable by simply increasing U.S. electric generation efficiency to just over 50 percent—by wasting only half the fuel. There is proof that this is possible. Great Britain opened its electric markets in 1989 and market competition, combined with increased use of natural gas generation, has reduced carbon dioxide from electric generation by 39 percent.

This translates to a 13 percent drop in total U.K. carbon dioxide emissions in just six years (U.K. electric prices have dropped by 15 percent to 20 percent.)

To mitigate climate change and save money, we must pick the low-hanging fruit by improving efficiency. We must deregulate electricity so market forces can achieve greenhouse gas reduction targets and lower electric prices. This will help the United States economy and help those developing nations that adopt similar efficiency standards. A global partnership of industry and governments has had success in eliminating ozone-depleting chemicals and markets have found very inexpensive ways to reduce U.S. sulfur emissions. We must find ways to make a global partnership work to reduce carbon dioxide.

We have discussed the idea with attendees at the second annual Wirth Chair meeting sponsored by the University of Colorado and gained perspective on how this approach compares with taxes, carbon caps, and renewable portfolio standards. Only time will tell if the approach makes it through the many necessary screens to become part of the administration's approach, or better still, new national law.

Rationale for a Federal Fossil Fuel Efficiency Standard

There is an accepted political wisdom that control of pollution is the legitimate purview of the federal government, and laws controlling emissions of oxides of nitrogen, sulfur, particulates, and other regulated pollutants apply to power plants in all sectors. Municipal plants, public power plants such as those owned by the Tennessee Valley Authority, REA plants, investor-owned utility plants, and independent power producers are all subject to federal pollution regulations, at least on new plants. Applying a fossil fuel standard to all power plants can be justified as a way to control a very important global pollutant—carbon dioxide. Although plants built before the original Clean Air Act was passed in 1972 are exempt from many of the act's provisions, there is precedent for applying new pollution-control standards to existing plants. The federal limitations on sulfur were applied to all of the largest power plants in 1992. The FFES is a way to cut through the morass of state, local, and federal rules that created the special interests that now oppose competition for their power.

Avoid a Federal Fuel Choice

It is an unfortunate feature of recent energy legislation that the federal government has felt it was able to discern the fuel of choice, or at least the fuel to avoid. On five occasions since 1960, the government has decided on the fuel of the future—each time selecting a different fuel than the last time. This has caused chaos, cost society unnecessary money and problems, and has proven repeatedly that selecting the "best" fuel should be left to self-correcting markets. Coal was the fuel of choice before 1960, but then oil looked better, and many coal plants were, under government prodding, converted to burn oil to lower the local sulfur and particulate emissions. Then came OPEC's price increases. Oil dependency and political instability resulted in a new focus on coal generation. Gas was regulated to death for thirty years, with prices held to eleven cents per thousand cubic feet at the well for years. Faced with these artificial prices, energy-exploration companies stopped drilling for gas. In time, the resulting gas scarcity led to banning the use of gas in power plants. Today, gas is abundant, thanks to deregulation and market-based prices, and many environmentalists would cheerfully mandate a switch to gas and away from coal.

It is particularly tempting to enact legislation forcing a switch from coal. Coal is nearly pure carbon, and all of its energy release results from combining (or combusting) carbon and oxygen to form carbon dioxide. By contrast natural gas is a hydrocarbon, consisting of four atoms of hydrogen and one of carbon. Roughly 40 percent of the energy release from natural gas combustion is from burning hydrogen to form H_2O, or water. From a carbon dioxide emissions point of view, it is better to burn gas. Oil is also a hydrocarbon, and burning oil releases only 75 percent of the carbon dioxide emissions associated with producing the same amount of energy with coal. Of course, both oil and gas are more expensive than coal, and the dominant energy resource in America is coal. Mandating a switch from coal would create huge political problems and would raise powerful opposition to decarbonization. If the Clinton administration proposes legislation to force a switch from coal to other fuels, Congress will probably balk. The result of attacking "King Coal" could be inaction. Attack inefficiency instead.

In fact, the attack on coal could also prove to be as misguided as earlier bans on natural gas use. We cannot predict market responses or new technology for the conversion of any fuel to energy in the future. Different companies, large and small, have dreams and all will develop programs to reach their dreams to supply energy. Each favors a different strategy or a different fuel. There is steady progress in technology to enhance the efficiency and economics of each fuel. Work has been underway in Wyoming for eight years to treat coal to make it cleaner and more energy efficient. A newly invented coal pulverizer grinds coal three times as fine as present technology, allowing sulfur to be removed more easily and the coal to burn more cleanly. Biomass-energy projects create a very hydrogen rich gas when they gasify cellulose, and this high hydrogen fuel requires expensive engine modifications. Blending the biomass gas with coal gas produces a fuel that can be burned in a standard engine. If government forces a business to choose one fuel, these advances will be curtailed. New laws should simply steer society toward efficiency and encourage long-term energy solutions that do not release any fossil carbon.

The Fossil Fuel Efficiency Standard avoids government selecting the fuel of choice and avoids the very messy job of measuring the atomic carbon content of each load of coal. This would be necessary if we adopted a carbon dioxide standard. By contrast, the energy content of all fuel is already routinely measured and is the basis for its price. When burning wood chips, cotton waste, bagasse, or rice hulls, the carbon dioxide produced is no greater than would be released if the biomass decayed naturally. With a Fossil Fuel Efficiency Standard, there is no counting of biomass fuel. Just count the megawatt-hours of power output. A carbon dioxide standard would force each power plant to sort out how much of its carbon emissions came from renewable fuels and how much from fossil fuel—a needless complication.

Fossil Fuel Efficiency Is Essential to a Sustainable Future

A sustainable future, with or without global warming concerns, must be based on renewable energy. We will exhaust the known fossil reserves in a few centuries, and even if there were to be vast new discoveries, the planet will run out of fossil fuel in the relatively near future at the present and projected rate of use. Good policy should encourage the energy-conversion industry to make efficient use of the fossil fuel burned and encourage a move toward economically viable renewable energy. Fully open markets would induce more efficient fuel conversion to energy. Ending all subsidies on fossil fuel would move the fuel-conversion industry toward more reliance on renewable energy sources. For the reasons cited above, it is unlikely that government will open up all electric markets or end all fossil fuel subsidies any time soon. Nonetheless, the government can guide the decisions of the energy industry in the direction that fully free markets would go, and the Fossil Fuel Efficiency Standard achieves this aim.

A Fossil Fuel Efficiency Standard Is Equitable and Predictable

A Fossil Fuel Efficiency Standard will give everyone a fair chance, whether they are an existing power company, a new power company, or a future power company. It will start to internalize the economic costs of burning fossil fuel, and will promote efficiency. It will act like a capital subsidy to the efficient technologies and a capital tax on the inefficient approaches. It will give everyone a clear road map showing where we want to go (i.e., to double present fossil efficiency), but will also give business enough time to adapt. The gradual tightening of this standard is very important economically. Power plants require added maintenance and capital investment every year. If the owners know that these plants will become increasingly expensive to operate ten years from now, due to the Fossil Fuel Efficiency Standard, they can plan an orderly extraction of the plant's value. This means the old power plants will be run as long as they can operate with minimal maintenance and no rehabilitation. Instead of spending money on extending the lives of inefficient plants, the power companies will put their money in new and doubly efficient plants that, incidentally, also are twenty times cleaner than what they replace, or will invest in heat recovery and increased efficiency in existing plants. These are all choices that can be guided.

This Fossil Fuel Efficiency Standard approach will signal a growing emphasis on renewable energy and will induce private investment in renewable research and development of renewable sources of energy, such as biomass, solar energy, wind, and tides, but will force renewables to compete with efficiency. This will drive renewable-energy prices down.

Entrepreneurs often try, to paraphrase Peter Drucker, "to take what is known and extrapolate its impact on the future." Entrepreneurs regularly stake out an early position before others see the future they see, so as to profit from the expected changes. Competition works because each entrepreneur has a different dream, and bets on a different advance of technology or a different vision. By providing a growing premium or reward for every technology and every approach to energy production that reduces the use of fossil fuel per unit of useful energy, government will send an unmistakable signal to all power entrepreneurs, who will pursue unimaginably diverse actions to position their companies to benefit from the announced future.

The consumer will reap benefits. Advances in fossil efficiency will be made in every area, and these advances will lower the costs of producing useful energy, carrying the promise of creating great wealth for the company that makes the advance. Then some other company will want some of the profits and will lower the prices they charge for energy, all to gain market share. Each act of the resulting play will lower the price of energy to the consumer. Society will also benefit from the resulting reduction in the amount of carbon dioxide and other pollutants that are produced. Besides mitigating global-climate change, the reduced pollution brings other benefits. There is great concern among foresters that America's trees are dying from acid rain even though sulfur has been curtailed. These actions will dramatically lower the NOx production that is contributing to the death of our forests.

Predicted CO_2 Reduction from Competition and a Fossil Fuel Efficiency Standard

Realistically, there will not be a complete opening of competition, but there will be movement. Every utility is preparing for the day when full competition arrives. There will be behavior changes to prepare for competition as states strive to improve the effectiveness of their regulation of power and partially open competition. There are efforts underway to eliminate or modify some of the barriers to efficiency. If the process is also guided by a Fossil Fuel Efficiency Standard, decarbonizing will move faster.

The chart in figure 17 looks at the impact of three scenarios on the emissions of carbon dioxide from the generation of all U.S. electricity over the next twenty years. It is based on estimates issued by the United States Department of Energy regarding the future generation of electricity. The

Fig. 17. CO$_2$ Emissions from U.S. Electric Power Generation[121]

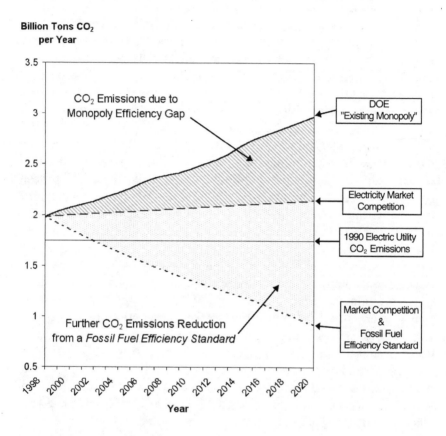

DOE forecasted 2.5 percent growth of electric use per year, and made a detailed projection of changes in the mix of generation plants due to technology. We have applied engineering constants to the fossil fuel that would be burned each year under the DOE forecast to estimate the metric tons of carbon dioxide emissions per year. Note that with "business as usual," the total emissions from electric production rise from roughly 2,000 billion metric tons to about 2,750 billion metric tons in twenty years.

The center line on the graph is our internal forecast of the CO$_2$ emissions that will actually result with the current and expected moves to deregulation and partial unleashing of market forces. We are projecting that the increases in fossil efficiency from replacing some old power plants with new technology will offset the 2.5 percent growth of electric usage, with the resultant CO_2 emissions held roughly constant. We can expect uncompetitive plants in California and Massachusetts to be either modified or retired as soon as the transition period passes (five to seven years). We also anticipate an increase in combined heat-and-power plants. There will be some penetration of microturbines that are 40 percent efficient and can recover another 30 percent of the fuel as heat for hot water. A wide variety of approaches will chip away at inefficiency.

The third projection, the lower line, is what we believe will happen if the move to competition is guided by the Fossil Fuel Efficiency Standard that yearly tightens the total fossil fuel allowed per megawatt-hour of total electricity. We have kept the DOE estimate of 2.5 percent growth of electric usage, even though a competitive power industry will begin installing end-use efficiency devices. The CO_2 emissions drop to less than half of present and fall far below the 1990 emissions. (Remember that the Kyoto Protocol calls for the United States to drop its greenhouse gas emissions to 7 percent below the 1990 levels by 2010.) The assumptions about the efficiency of the United States electric generating industry in this projection are very simple. The projection simply assumes that the industry can reach an average efficiency equal to what Trigen Energy Corporation was achieving in eighteen separate power plants in 1998. With twenty years to change, and technology being improved continuously, this does not seem like an impossible dream. Each of the plants Trigen has built save money for the users and earn a return on investment, so it would seem that this level of fossil fuel efficiency could be obtained with savings to the economy.

What is most important is two shaded areas between the DOE projections and the projections we have made with a Fossil Fuel Efficiency Standard. The cross-hatched area between DOE "business-as-usual" projections and the electric market with competition is the CO_2 emissions reductions that come from the *Monopoly Efficiency Gap*. This is the carbon dioxide that is attributable to the needless burning of fuel with business as usual—protected monopoly. The shaded area with dots is the further CO_2 emission savings that would result from a Fossil Fuel Efficiency Standard. The lower projection can be achieved if we deregulate all electricity, end all barriers to efficiency, and end all fossil fuel subsidies and guide power entrepreneurs to less fossil dependence. The Fossil Fuel Efficiency Standard can guide behavior and avert the "tragedy of the commons."

Both actions will call forth a burst of new technology and will increase the demand for efficient power plants and combined heat-and-power plants, which will lower their installation cost. This new and more affordable technology will successfully compete with older, less efficient approaches all over the world and will have an impact on mitigating climate change far beyond U.S. borders. Since the Fossil Fuel Efficiency Standard inexorably tightens the allowable fossil fuel per megawatt-hour, all power entrepreneurs must eventually embrace renewable energy. This will induce a race to make renewable-energy technologies competitive with fossil fuel technologies. Again, the winning technologies will find a ready home all over the world, and we will be on the way to mitigating climate change without disrupting any country's economy.

Fossil Fuel Efficiency Standard Versus the Alternatives

Various people who have reviewed the prepublication versions of this book have asked the same question. "Tom," they say, "aren't you inconsistent, calling for an end to government monopoly regulation and elimination of government rules that are barriers to efficiency, and then calling for a new government intervention in the economy, a Fossil Fuel Efficiency Standard?" The answer is a limited yes, based on both pragmatism about what may be possible to solve the climate problem, and a deeper belief that government has an essential role to play in helping society avoid the excesses of self-interest.

There are simply many examples in our modern society where the actions of individuals that meet their self-interest, even meet the individual's enlightened self-interest, result in harm to the society. Garrett Hardin used the example of cows grazing on the community common, but there are other poignant examples. Peter Bradford and I were in India at a conference on electricity in Jaipur, and we were driven across town one evening to a banquet. As we approached a train crossing, the gates came down, stopping all traffic. In countries where the government enforces rules to keep all cars in one lane, cars wait on both sides of the track in their respective lanes and move immediately when the train passes. Not so in Jaipur. Staying in one's lane is not enforced. As the train gates came down, our driver and several others quickly pulled out into the righthand lane (Indian cars drive on the left), passing the other cars to get closer to the train gate. The same game was being played on the opposite side of the track.

Peter noted that we had a great example of the "tragedy of the commons." If our driver waited in his lane, competing drivers would fill the other lane anyway; it was in our driver's self-interest to jockey for position. However, when the gate opened, two lanes of cars that were headed south were nose-to-nose with two lanes of cars that were headed north. No one moved. Those who eschew all government intervention should have been there. Being obliged to stay in your lane averts problems for everyone.

Global climate change is a comparable problem in need of government leadership. There is a clear gain to society from guiding everyone toward less reliance on fossil fuel. If done thoughtfully, it will not disrupt the economy but will increase our general well being. A proper lane rule, evenly enforced, will cause all drivers to reach their destinations sooner. A properly designed and enforced Fossil Fuel Efficiency Standard will lead to a balance between lower energy costs and a more healthy environment, and will buy time to move to a fully sustainable global energy infrastructure. It is not inconsistent to object to government regulations that are based on conditions that no longer exist, or have been shown to work badly, but to continue to believe that government intervention is necessary to guide society.

EXPECTED ELECTRIC COST REDUCTIONS

This book's introduction began by proclaiming that we had found the secret word—competition. Our thesis throughout has been that mitigation of climate change and decarbonization of the economy will occur and will save money if market forces are unleashed and barriers to efficiency are eliminated. The real price reductions in other regulated industries have ranged from 27 percent to 58 percent ten years after regulations were eased. Our prediction for the electric industry is a 30 percent to 40 percent real-price drop in ten years after deregulation, which would surely boost the economy.

Industry and economic pundits usually expect prices to rise with inflation, yet since the start of the industrial revolution, real prices have fallen. The standard of living that people enjoy today is a result of falling real prices. Each person's labor buys more and more goods and services over time. The driver of this growing standard of living is always the same—competition.

While the scientific method gets the credit for the rapid advance of knowledge, and much of the drop in costs over time can be traced to technological advances, it is competition that drives the search for the deployment of technology. Competition is the secret word.

Let's look at one more industry, in this case petroleum. Competition has been present during most of the time since discovery of oil in 1859, although the industry moguls have done their best to shut out competition. In 1973, the Organization of Petroleum Exporting Countries (OPEC) made a new run at shutting out competition and used their cartel powers to restrict oil output and raise prices. This held back competition for thirteen years. A period of rampant inflation ensued through the late 1970s and early 1980s. Oil prices were driven up to $30 per barrel.

In 1986, the demand for oil collapsed, and prices began to fall. The OPEC cartel could not hold production back. Industry and observers predicted dire consequences for the oil industry, as competition suddenly became rampant. Producers once again had to fight over sales, and had to drop their prices to make sales. So what happened? Was there a disaster? Looking at the robust economy of 1998, it is hard to find the disaster.

Daniel Yergin, winner of the 1992 Pulitzer Prize for his book about the oil industry, titled *The Prize,* gave this account in 1998: "One of the things that strikes me about this (the oil) industry is that it can, at $15 or $16 a barrel, do things that it thought it couldn't do at $30 a barrel a decade ago. It's an industry that is being transformed by technology and computers. It's an industry that can do much better at lower prices. It's an industry that surprised itself."[122]

These comments may be applied, in advance, to the electric industry, which has not experienced true competition for ninety years. The electric industry will be able to sell at a 30 to 40 percent discount and do things it cannot do today at full current prices. The electric industry will be transformed by technology and computers. The electric industry can do much better at lower prices. The electric industry will surprise itself.

10

ERRATA—
Energy Regulatory Reform
and Tax Act

The dictionary defines "errata" as errors in printing or writing, especially such errors noted in a list of corrections and bound into a book.[123] What is needed to improve the economy, clean up the environment, and decarbonize is a comprehensive federal act that fixes all of the errors in existing legislation with respect to the regulation and taxation of energy. This comprehensive act can be titled ERRATA, or the Energy Regulatory Reform and Tax Act. Here is a summary of the proposed ERRATA, which is, in fact, a summary of the ideas presented throughout this book.

(1) Deregulate all electric generation and sales and modernize regulatory laws impacting energy

- Allow all generators to parallel with the grid, to purchase backup power from any source, and to sell at retail to any buyer.

- Remove prohibitions against private distribution of electric power.

- Ban states from restricting the generation and sale of power by any entity.

- Pay stranded costs of existing utilities only if they auction their generating assets and power-purchase contracts to determine the market value, then pay the difference between book value and sale prices. Charge all users of power a transition fee over five years to recover the stranded costs, but exempt from transition charge power from combined heat-and-power plants with over 50 percent efficiency or that use renewable energy.

- Establish an independent body to set national interconnection standards.

- Repeal the Public Utility Holding Company Act and much of the Public Utility Regulatory Policy Act.

(2) Modernize environmental regulation of power plants

- Set performance standards for all criteria pollutants per megawatt-hour of heat and power produced that apply to all generation companies' average annual production. Schedule allowed pollutant per megawatt-hour to decline over twenty years.

- Set a new Fossil Fuel Efficiency Standard that applies to all generation companies' average annual production, with allowable fossil fuel per megawatt-hour scheduled to decline over twenty years.

- Allow trading to achieve pollution allowance standards, and establish fines for all excess pollution, after the effect of trading.

- End New Source Performance Standards for power plants. (The objectives of NSPS are completely achieved by performance standards that tighten over time, and this takes the government out of individual plant regulation.)

- End all requirements to obtain air permits for new power plants, relying on pollution allowances to control emissions.

(3) Change taxation to encourage efficiency.

- Set tax depreciation lives on all power-production equipment based on size of engine or plant, regardless of efficiency.

- Encourage district heating with tax credits for investment in new piping systems to move thermal energy between multiple buildings.

- Provide a tax credit to the purchaser of any major appliance, automobile, or truck that is certified to be 50 percent more efficient than the average of all comparable appliances or vehicles in use the prior year. The DOE would adjust the efficiency averages every year and certify the models that achieve the 50 percent efficiency gain.

(4) Include energy efficiency in all federal activity and funding.

- Mandate standards of energy use per square foot in all federal facilities and schedule the standards to tighten over twenty years.

- Add a condition to all federal support of research and education that the receiving institution achieve a standard of energy use per square foot of space and schedule the standards to tighten over twenty years.

- Allow all federally supported installations including public housing, military bases, Veterans' hospitals, and national laboratories to keep 50 percent of any energy savings for use by local administration as they see fit to encourage conservation.

- Increase support of the Energy Information Agency to collect the data needed to establish, monitor, and modernize the performance standards and efficiency rules.

(5) Take miscellaneous federal actions to promote energy conservation.

- End all direct federal subsidies of fossil fuel exploration and production.

- Establish national building codes that require all new commercial buildings of some specified size to be district energy compliant (i.e., to use central hot water or steam to carry heat to individual spaces), require occupancy sensors to control lighting and computers by a date certain in all buildings, and establish new construction insulation standards that increase over twenty years.

- Establish a comprehensive national shade-tree-planting program with a goal of shading every building of three stories or less and every mobile home with deciduous trees.

- Amend antitrust statutes to specifically include all energy production, distribution, and sales under the existing antitrust rules and fund vigorous enforcement to ensure a competitive market for energy production and services.

- Ensure siting reviews are appropriate for smaller CHP plants and distributed electricity generation.

The passage of ERRATA will have consequences far beyond the federal actions specifically mandated. It will become clear to state governments that their own laws are standing in the way of improved efficiency at medical centers, universities, and other facilities and thus may limit federal support. This will cause enlightened state governments to modernize their own rules that are barriers to efficiency. This will stimulate the passage of an ERRATA in every state that addresses the unique problems of outmoded laws in that state. The mere introduction of the legislation will send signals to every energy firm, energy-efficiency company, building developer and designer, equipment manufacturer, and consultant that the future will contain rewards for improving efficiency and rewards for reducing dependence on fossil fuel.

The passage of ERRATA will allow environmental agencies to achieve their air pollution reduction goals much more efficiently, and will free the staff people that now examine each new power plant application. These environmental professionals can be redeployed to other environmental management tasks, but will be heavily recruited by the energy industry, which will need to rebuild most old power plants. This modernization of environmental regulation will remove much of the command and control approach and will unleash market forces to achieve the tightening performance standards. The pollution performance standard provision of ERRATA alone will cause the U.S. power generation fleet to be completely rebuilt over the next two decades, spurring an investment of $200 to $400 billion. Steam fitters, pipe fitters, construction trades, engineering and construction firms, equipment manufacturers, lending institutions, and heat and power developers will all see a sustained boom in activity.

Because the energy industry is so large, it impacts the cost of virtually every good and service produced in the United States. The passage of ERRATA will lead to a drop in electric prices of 30 to 40 percent over the ensuing decade and this will lower the cost of making all goods and

services. This will then lead to increased competitiveness of U.S.-pro-
duced products and a consequent surge in U.S. exports. At the same
time, the reduced demand for fossil fuel will cut U.S. imports of oil and
thus improve the U.S. balance of payments. If other countries do not
emulate ERRATA, the United States will gain a tremendous competitive
advantage. But no country can afford to stand by with outmoded regulations and barriers to efficiency and watch its market for goods evaporate. Other countries will act and enact their own ERRATA. By taking a
leadership role, and leading by example, the U.S. will positively impact
the energy efficiency of every other country.

The great debate about the wisdom of signing the Kyoto Protocol
becomes largely irrelevant if the United States passes ERRATA. When the
strongest economy in the world modernizes its approach to energy and
stimulates efficiency and lessened dependence on fossil fuel, *competition,*
the secret word, will come into play internationally. Global competition
will drive each nation, developing or developed, socialist or democratic,
to clean up its own act and to lower the cost of its energy production and
distribution. Outmoded laws have held back energy-efficiency technology
all over the world for four decades and have kept power-generation efficiency stagnant while every other industry gained productivity every year.
When ERRATA is passed and unleashes pent-up technology, market
forces will achieve and exceed the greenhouse gas reduction goals of the
Kyoto treaty. Global climate change will be mitigated as a by-product of
improving the economic efficiency of energy conversion and use.

We began this book by explaining the scientific basis for concern that
human actions are having a discernable effect on global climate. The challenge, in June 1997, of then Secretary of Energy Federico Peña to suggest
programs to mitigate climate change that would not disrupt the economy
inspired a year's effort to pull together all of the lessons from a career
of trying to change the way the world makes power. Ironically, all of the
actions proposed should be taken for economic reasons alone. The proposed actions should be supported by the most adamant doubter of climate change as well as by those, like me, who believe climate change is
the most pressing global problem we have ever faced. There is no need
to delay improving our economy, our air quality, and our standard of
living while we wait to gain more certainty about climate change. The
mitigation of climate change is simply the icing on a very large cake.

Epilogue

In the five months since *Turning Off the Heat* was released, several noteworthy actions have been taken by federal and state governments as they recognize the need for increasing energy efficiency, eliminating barriers to efficiency, and unleashing market forces. Here's a brief update:

1. The Department of Energy held a "CHP Summit" on December 1, 1998, in conjunction with the newly formed U.S. Combined Heat and Power Association (USCHPA). Assistant Secretary of Energy Dan Reicher announced a CHP challenge, with a goal of doubling by the year 2010 the percent of power generated in the United States by combined heat-and-power plants. Roughly 9 percent of America's power was generated in plants that recovered some or all of the normally wasted heat in 1997. Department of Energy officials calculate that we need to build 46,000 megawatts of new CHP plants to meet the goal. This would require an investment in energy infrastructure of $30 to $50 billion over ten years.

The goal has been stated as follows:

"CHP Challenge Addresses Barriers to CHP Development"

The Department of Energy seeks to open a national dialogue on CHP technologies to raise awareness of the energy, environmental and economic benefits of CHP, and promote innovative thinking about ways to accelerate the use of CHP.[124]

2. The Environmental Protection Agency has issued a final rule calling for reductions in nitrous oxide emissions from existing power plants in twenty-two states and the District of Columbia.[125] For the first time, the EPA is permitting the states to allocate emission allowances among power plant and industrial boiler owners on an output basis. Broad use of output-based standards (recommended on page 179 of the present volume) will enable power-plant operators to comply with the new rules by recovering heat while displacing other boiler emissions. Senior EPA officials have publicly admitted that the current approach to air-quality regulation is environmentally and economically "dysfunctional" and have expressed a willingness to explore regulatory approaches to air quality that encourage energy efficiency.

3. The Clinton administration's revenue proposals, issued in February 1999, request an 8 percent investment tax credit for new combined heat-and-power installations.[126] To qualify, a plant of less than fifty megawatts of electric output must achieve over 60 percent total efficiency, just short of twice the national average efficiency for electric production. If the plant is over fifty megawatts of electric output, the CHP facility must achieve 70 percent efficiency. Congress has not yet acted on these proposals.

4. The Department of Energy has commissioned a study of present electrical interconnection rules throughout the United States. These are the rules issued by each monopoly utility governing what protective devises must be installed to allow a nonutility generator to be connected to the electric grid. The department believes that unnecessary interconnection requirements impede deployment of a vast array of proven distributed power technologies. Informed sources project that the administration's

deregulation bill will call for the Federal Energy Regulatory Commission to be given power to set national interconnection rules.

5. On October 27, 1998, the EPA presented Trigen Energy Corporation with an international "Climate Protection Award," citing the actions of Trigen in 1997 that reduced CO_2 emissions by over 1.2 million tons. Other corporations honored included IBM for a chip-making process that avoided the use of a greenhouse gas; McDonald's for innovative energy-saving restaurant design; automobile manufacturer Toyota for its Prius hybrid vehicle that uses a small gasoline engine combined with an electric motor and battery to power the car; and several other companies.

6. Several major U.S. companies, including Millennium Inorganic Chemicals, Equistar, Monsanto, and others, all recently announced decisions to deploy CHP and other energy-saving technologies to reduce CO_2 emissions and save money.

7. On pages 175–76 of this volume, I briefly mention a dispute between Trigen and a utility that had erected numerous inappropriate barriers to competition and to the deployment of efficient generation. In December of 1998, a jury agreed with our antitrust and tort claims and awarded Trigen more than $32.5 million in damages, sending a clean pro-competitive message to the energy industry. I do not believe that the inevitable appeal process will dilute that message.

Notes

1. Alexander Deconde, *This Affair of Louisiana* (New York: Charles Scribner's Sons, 1976), pp. 178–79.

2. "Energy Primer in the Working Group II SAR", Global energy-related CO_2 emissions by major world region in Gt C/yr.

3. "Summary for Policy Makers: The Science of Climate Change—IPCC Working Group I." Intergovernmental Panel on Climate Change (IPCC), 1995, Section 4, Climate Change 1995: IPCC Second Assessment Report. Found at http://www.ipcc.ch/cc95/wql/htm, page 3 of 6.

4. D. James Baker, Undersecretary of Commerce for oceans and atmosphere, April 10, 1998, NOAA website http:///www.publicaffairs.noaa.gov.

5. J. M. Barnola, D. Raynaud, C. Lorius, Y. S. Korotkevich, *Historical Record from the Vostock Ice Core Data,* September 1994, found at http://www.daflight.demon.co.uk/science/index.htm.

6. Ibid.

7. "The Vostock Ice Core Data," http//www.daflight.demon.co.uk/science/index.htm.

8. *Wall Street Journal*, April 16, 1998.

9. "Millennial-scale Climate Instability During the Early Pleistocene Epoch"(letter), *Nature*, April 16, 1998.

10. Population Information Network (POP!N) Gopher of the United Nations Population Division, Department for Economic and Social Information and Policy Analysis.

11. Elenor M. Fox, "Eastman Kodak Company vs. Image Technical Services, Inc.: Information Failure as Soulor Hook?" *Anti-trust Law Journal* (Spring 1994): 759–67.

12. Robert Crandall and Jerry Ellig, *Economic Deregulation and Customer Choice: Lessons for the Electric Industry* (Fairfax, Va.: Center for Market Processes, 1997).

13. Theodore J. Lowi, *The End of Liberalism: The Second Republic of the United States* (New York: W. W. Norton and Company, 1979).

14. *Electric Utility Restructuring: A Guide to the Competitive Era: Chapter 1: Origins of the Modern Deregulation Debate,* Box 1–3.

15. Note that this is not a universal truth. In Cleveland, Ohio, there are dual wires, one from an investor-owned electric utility, Centerior, and one from a municipal utility owned by the city of Cleveland.

16. Tina Kaarsberg and R. Neal Elliott, "Combined Heat and Power: Saving Energy and the Environment," *Northeast-Midwest Economic Review* (March/April 1998): 4.

17. Modified from Evertt M. Rodgers, *Diffusions of Innovations*, 4th ed. (The Free Press, 1995), p. 11.

18. Bell Labs Museum.www.lucent.com/museum/1947.html.

19. Testimony of Captain R. Michael Baiada, RMB Associates, before the National Civic Aviation Review Commission, October 8, 1997, p. 2.

20. "Supply of Oil," *Scientific American* 12 (June 27, 1857): 1.

21. Microsoft Bookshelf '95, Chronology, 1859: "Petroleum production begins at Titusville."

22. James Trager licenses the *People's Chronology* (Microsoft Bookshelf CD-Office 97) from Henry Holt and Company, Inc. (1994).

23. *American Heritage Dictionary of the English Language*, 3d ed. (Houghton Mifflin Company, 1992). Electronic version licensed from InfoSoft International, Inc.

24. Even with today's technology, it is difficult to make a rotating direct current generator very large.

25. Jeffery A. Hayes, *Tesla's Engine: A New Dimension for Power* (Wisconsin: Tesla Engine Builders Association, 1994).

26. Microsoft Bookshelf '95, Book of Quotations.

27. www.geocities.com/wallstreet/flour/3748/insul.

28. Governor Pataki: "New Era of Rate Relief Launched on Long Island," Long Island Power Authority press release, June 1, 1998.

29. Charles Bayless, "Less Is More: Why Gas Turbines Will Transform Electric Utilities," *Public Utilities Fortnightly*, December 1, 1994.

30. "Staff Paper Discussing Commission Responsibilities to Establish Rules Regarding Rates and Exemptions for Qualifying Cogeneration and Small Power Production Facilities Pursuant to Section 210 of the Public Utility Regulatory Policies Act of 1978," *Federal Register* 44, no. 129 (July 3, 1979): 38863.

31. *American Heritage Dictionary of the English Language.*

32. *Cogeneration Planner's Handbook*, Figures 2–23 (slightly modified), p. 56.

33. Steam contains two kinds of energy: the latent heat of vaporization or quantity of heat necessary to change the liquid to vapor without change of temperature, and the additional energy in the superheated vapor. Steam turbines with buckets remove most of the superheated vapor energy, but do not convert the latent heat of vaporization to energy, and this imposes the limit on efficiency of conversion. Nikola Tesla invented a bladeless steam turbine that was demonstrated but never commercialized, and he claimed that his turbine could achieve upward of 97 percent efficiency by recovering the latent heat of vaporization, if properly staged. Tesla was intensely secretive and never wrote how to achieve such efficiencies. A group of Tesla fans have built various working models of the bladeless turbines and have an association, Tesla Engine Builders Association, 5464 N. Port Washington Road, Suite 293, Milwaukee, WI 53217. See Hayes, *Tesla's Engine.*

34. Photograph taken by Mark Parchman, Parchman Photography, Craig, CO 81625.

35. Created from the author's own experiences and collected data from U.S. DOE/EIA.

36. Bayless, "Less Is More."

37. We estimate later that competition will lower electric prices by 30 to 40 percent in ten years after regulations are eased, based on experience deregulating five other industries. Current total electric charges in the United States are roughly $210 billion per year.

38. Charles E. Little, *The Dying of the Trees* (Penguin Books, 1995).

39. *World Almanac 1995* and American Automobile Manufacturers Association press release, "World Motor Vehicle Production Hits Record High in 1997," February 24, 1998.

40. Diesel generator sets achieved 32 to 38 percent efficiency by 1975 and produced power at the point of use, at the needed voltage. Central electric plants over the United States averaged a delivered efficiency of 29.5 percent. Internal Trigen Energy Co. studies.

41. Personal recollection from the hearings in 1979 in Albany, New York.

42. *Concise Columbia Encyclopedia* (Columbia University Press, 1995).

43. *American Heritage Dictionary of the English Language.*

44. Ibid.

45. David C. McCullough, *The Path Between The Seas: The Creation of the Panama Canal* (Simon and Shuster, 1978).

46. *American Heritage Dictionary of the English Language.*

47. Ross Gelbspan, *The Heat Is On: The High Stakes Battle Over Earth's Threatened Climate* (Addison-Wesley Publishing Company, Inc., 1997), p. 175.

48. Fred Cottrell, *Energy and Society* (New York: McGraw-Hill, 1955), p. 2.

49. The *People's Chronology,* licensed from Henry Holt and Company.

50. Cottrell, *Energy and Society,* p. 6.

51. For further information, see William H. Calvin, "The Great Climate Flip-flop," *Atlantic Monthly* 281, no. 1 (January 1998): 47–64.

52. "Climate Change: State of Knowledge," Executive Office of the President, Office of Science and Technology Policy, October 1997, pp. 7–8.

53. Little, *The Dying of the Trees.*

54. Ibid.

55. "Pothole in the Ozone Layer," *Washington Post,* March 15, 1997.

56. Comprehensive Electricity Competition Act, drafted by the U.S. Department of Energy and released in June 1998.

57. "Kyoto Protocol Opens for Signature . . . ," *Environment Reporter* 28, no. 45, March 20, 1998.

58. Shell Corporate Web Page: http://www. shell. com. "Shell Leaves Coalition that Opposes Global Warming Treaty," *Washington Post,* April 22, 1998.

59. Freeman Dyson, *Imagined Worlds* (Harvard University Press, 1997).

60. Garrett Hardin, "The Tragedy of the Commons," *Science* 162 (1968): 1243–48.

61. Adam Smith, *An Inquiry into the Nature and Causes of the Wealth of Nations* (University of Chicago, 1976).

62. Fredrich Von Hayek wrote several books and articles in which this sentiment was expressed, including "The Use of Knowledge in Society," *American Economic Review* 35 (September 1945): 519–30.

63. Personal correspondence from Mike Thompson.

64. *Ecolabeling Criteria for Washing Machines,* U.K. Ecolabeling Board, August 1992.

65. Author's personal notes, October 6, 1997.

66. Miriam P. Ye and R. Neal Elliot, *Energy, Efficiency, Pollution Prevention, and the Bottom Line,* American Council on Energy-Efficient Economy, August 1997.

67. Ludwig Von Mises, *Human Action: A Treatise on Economics,* 3d ed. (Chicago: Henry Regency Company, 1966), p. 3.

68. The Vanguard Group and U.S. EPA.

69. "Cleaner Power," *Economist,* April 18, 1998, p. 17.

70. Hardin, "The Tragedy of the Commons."

71. Jeffrey L. Cruikshank and David B. Sicilia, *The Engine That Could* (Boston: Harvard Business School Press, 1997), pp. 34–35.

72. Ibid.

73. Ibid, p. 117.

74. U.S. EPA, *Characterization of Municipal Solid Waste in the United States: 1995 Update,* p. 117.

75. Modified from Rodgers, *Diffusions of Innovations*, p. 23.

76. David G. Jefferies, IEE Presidential Inaugural Address. Paper on which speech in Boston was based, www.iee.org.uk/lectures/inaug97.htm.

77. Personal conversation with Sally Hunt, June 1996.

78. Peter Fox-Penner, *Electric Utility Restructuring: A Guide to the Competitive Era* (Peter Fox-Penner, 1997), p. 218.

79. "Blair Halts Gas Projects," *Independent Energy*, January–February 1998, p. 24.

80. It is common practice in the United States to measure the energy content of fuel and heat in British thermal units or BTUs, to sell steam in units of the pounds of water contained in the steam and to measure electricity in megawatt-hours. (1000 kilowatt-hours = one megawatt-hour). All forms of energy can be converted to megawatt-hours, and this is the common practice in most of the rest of the world. One megawatt-hour equals 3.412 million BTUs. The energy content of a pound of steam depends on the temperature and pressure of the steam, and can range from around 1,100 BTUs per pound to over 1,600 BTUs per pound. All of these measurements and conversions are well known to power-plant operators, and the government already collects all of the essential data to construct tables of fossil fuel efficiency.

81. Chart based on data from U.S. DOE Energy Information Agency, "Energy Info Disc," vol. 2, no. 3.

82. Ibid.

83. *Electricity Journal*, January–February 1998, pp. 56–63.

84. "When Virtue Pays a Premium," *Economist*, April 18, 1998, p. 57.

85. Personal communication with author, December 1997.

86. "Emission Reductions to 3.5 ppm NOx," *Diesel and Gas Turbine Worldwide*, January–February 1997, pp. 44–45.

87. Author's conversations with Dennis Orwig, president, Catalytica Combustion Systems, April 1998.

88. Personal communications with Joel Bluestein of Energy and Environmental Analysis, Inc., with Dennis Orwig, Catalytica Combustion Systems, and from "Emissions Reduction to 3.5 ppm NOx."

89. Ibid.

90. Climate Change State of Knowledge, Executive Office of the President, Office of Science and Technology Policy, P. S.

91. "Satellite Images Show Chunk of Broken Antarctic Ice Shelf," at http://www.eurekalert.org/releases/brkantartice.html. Theodore.Scambos@colorado.edu, press release, 4/18/98, University of Colorado at Boulder.

92. Mark Heartsgaard, "Severe Weather Warming," *New York Times*, August 2, 1998, pp. 48–49.

93. Ibid.

94. W. N. Keepin and G. Kats, "Greenhouse Warming: Comparative

Analysis of Nuclear and Efficiency Abatement Strategies," *Energy Policy* 16, no. 6 (December 1988): 538–61; RMI Publication #E88-29, www.rmi.org.

95. Rolf Stalebrant, "District Heating in Sweden: Great Potential to Replace Electric Heating," *District Energy* 83, no. 3 (1998): 8.

96. Petroleum refining involves the separation of a complex mix of chemicals based on their boiling points. All crude oil must therefore be evaporated before it is in a usable form.

97. When a liquid like oil or water is converted to a vapor like steam, physicists term the process a *phase change*, denoting the change from liquid to gas. Liquids require a significant input of heat to change phase. For example, heating water on your stovetop from the tap-water temperature of, say, 62 degrees F to 212 degrees F requires 150 units of energy. To then convert all of the 212 degree F water into 212 degree F steam requires roughly 1000 units of energy. When the steam cools, it condenses back to water, but must first give up the 1000 units of energy. The same is true of every liquid, though the proportions of energy vary with the chemical composition of the liquid.

98. Internal Trigen Energy Corp. data on performance of Grays Ferry cogeneration plant.

99. We worked for months with one state university that wanted to spend $65 million of their taxpayers' money on adding capacity to the campus energy system. We found ways to spend only $25 million and significantly increase the efficiency of production and use, but the utilities manager for the university was relied upon for the decision and he decided to keep us and other bidders out of his life, fearing job loss or uncomfortable personal change. The taxpayer money is now being spent on an inferior system. Sadly, there are many such examples because energy efficiency is simply not a priority for the provost of a typical university or president of a typical medical center.

100. Mark Kosmo, *Money to Burn? The High Costs of Energy Subsidies* (Washington, D.C.: World Resources Institute, 1987).

101. *Saving Energy the Natural Way*, consumer information from the California Energy Commission, www.energy.ca.gov/consumer/home/tree.html.

102. One company is building a new coal plant, but it is a special situation, located over a large coal bed.

103. Communication with Sharon Belanger, AES Corporation, March 19, 1998.

104. A commercial medium-speed dual-fuel diesel engine like the Model LSVB20, made by Cooper Industries, in this time frame produced 600 parts per million of NOx. In 1997, Catalytica completed 1,000 hours of testing on a small Kawasaki gas turbine with their "Xonon" combustor and announced under five parts per million of NOx in the exhaust, more than a 99 percent decrease in emissions. Press release, Catalytica, November 1977.

105. Clyde Wayne Crews Jr., "Electricity Reform: The Free Market Alternative to Open Access," *Electricity Journal* 10, no. 10 (December 1997): 35.

106. The Executive Committee, 1998 Calendar.

107. The actual PURPA rules are more complex, requiring heat output of at least 10 percent of energy in fuel, for which one-half credit is given, and added to electricity produced. The total is then divided by total fuel energy. The result must equal, or exceed, 42.5 percent. This means 45 percent in the minimum qualifying efficiency, while average U.S. efficiency is 33 percent from all generation. In other words, a qualified facility must be almost 150 percent of U.S. average efficiency.

108. Note that the reverse is true for gas-distribution companies. They experience peak loads in winter for heat, and would flatten their loads by encouraging installation of gas-turbine plants to generate electricity. The gas-distribution companies however, have seldom offered cogeneration inducement rates, perhaps because there are so many barriers.

109. Massachusetts General Law, chap. 146, secs. 46 and 48, "The Commonwealth of Massachusetts."

110. Tina Kaarsberg and R. Neal Elliott, "Combining Heat and Power: Saving Energy and the Environment," *Northeast Midwest Institute, Economic Review* (March/April 1998): 4, at http://www.nemw.org~periodic.htm.

111. Crandal and Ellig, *Economic Deregulation and Customer Choice.*

112. This chapter is a refinement of a paper titled "Barriers to Energy Efficiency" written by Mark Hall and Thomas R. Casten in October 1997 in response to a DOE request after a meeting on the same subject. That paper was singled out by Common Purpose, a not-for-profit organization committed to the environment, as the outstanding policy suggestion of 1997.

113. *Fossil Fuel Subsidies: A Taxpayer Perspective*, Taxpayers For Common Sense, www.taxpayer.net/fuelsubfact.htm.

114. Adam D. Thierer, *Energizing America: A Blueprint for Deregulating the Electricity Market*, The Heritage Foundation, Backgrounder no. 1100, January 23, 1997.

115. Charles Bayless, "Public Power—Time's Up," *Public Utilities Fortnight,* July 1, 1998.

116. A sales tax on electricity, regardless of supplier, would drive people to produce their own power to avoid the tax.

117. *Price It Right*, Alliance to Save Energy pamphlet, January 1998.

118. Social-cost pricing is implemented in this study by levying taxes (as a percent of fuel price) on fossil fuels equal to the estimated or measured environmental and health costs not already reflected in prices through environmental regulation. The shift in taxation is done on a revenue-neutral basis; that is, by lowering, or shifting, other taxes dollar-for-dollar for the taxes levied on

energy. Implementing social-cost pricing of energy confronts every producer and consumer with the correct costs, private and social, of using energy. Through millions of decisions daily, consumers and businesses buy and install energy-efficient equipment and processes faster and at less cost than any system of central control and planning could every achieve.

119. Breakfast in Denver, Colorado, during meeting of G-7 nations hosted by the DOE, June 20, 1997.

120. The Op-Ed is reprinted in its entirety here, by permission of the authors.

121. Nineteen ninety-six Annual Energy Outlook, Energy Information Administration, supplemented with internal Trigen Analysis.

122. Daniel Yergin, "Seeing More Mergers as an Industry Adjusts," *New York Times,* August 16, 1997, p. 10, sec. 3.

123. *American Heritage Dictionary of the English Language.*

124. http://www.oit.doe.gov/.

125. http://www.epa.gov/airprogm/oar/oaqps/airlinks/noxsipf.html.

126. "General Explanations of the Administration's Revenue Proposals," issued by the Department of the Treasury, February 1999. http://www.ustreas.gov/press/release/grnbk99.htm.

Glossary

Acid Rain. Also known as "acid deposition." Acidic aerosols in the atmosphere are removed from the atmosphere by wet deposition (rain, snow, fog) or dry deposition (particles sticking to vegetation). Acidic aerosols are present in the atmosphere primarily due to **emissions** of gaseous sulfur oxides (sulfur dioxide) and nitrogen oxides from both anthropogenic (burning **fossil fuel**) and natural sources. In the atmosphere these gases combine with water to form acids.

Aerosols. Particles of matter, solid or liquid, larger than a molecule but small enough to remain suspended in the atmosphere. Natural sources include particles from sea spray and clay particles as a result of weathering of rocks, both of which are carried upward by the wind. Aerosols can also originate as a result of human activities (burning **fossil fuel**) and in this case are often considered **pollutants**.

Albedo. The ratio of reflected to incident light; albedo can be expressed as either a percentage or a fraction of 1. Snow-covered areas have a high albedo (up to 0.9 or 90 percent) due to their white color, while vegetation has a low albedo (generally about 0.1 or 10 percent) due to the dark color and light absorbed for **photosynthesis**. Clouds have an intermediate albedo and are the most important contributor to the earth's albedo. The earth's aggregate albedo is approximately 0.3.

Atmosphere. The mixture of gases surrounding the earth. The earth's atmosphere consists of about 79.1 percent nitrogen (by volume), 20.9 percent oxygen, 0.036 percent **carbon dioxide**, and trace amounts of other gases. The atmosphere can be divided into a number of layers according to its mixing or chemical characteristics, generally determined by its thermal properties (temperature). The layer nearest the earth is the *troposphere*, which reaches up to an altitude of about 8 km (about 5 miles) in the polar regions and up to 17 km (nearly 11 miles) above the equator. The *stratosphere*, which reaches to an altitude of about 50 km (31 miles), lies atop the troposphere. The *mesosphere,* which extends up to 80–90 km, is above the stratosphere, and finally, the *thermosphere*, or *ionosphere*, gradually diminishes and forms a fuzzy border with outer space. There is relatively little mixing of gases between layers.

Backpressure Steam Turbines. A power-producing device in which one or more rows of blades on a shaft are pushed by the pressure of incoming steam. In the process, some of the incoming steam's pressure and temperature is converted very efficiently (above 97 percent) to mechanical energy which is used to drive an electric generator, or to drive a fan, pump, or other rotating machine. In a backpressure steam turbine, the steam leaving the **turbine** maintains enough energy for other purposes.

Barriers to Efficiency. A term coined by the author to describe rules, laws, regulations, and other practices that unintentionally restrict the implementation of efficiency improvement projects.

Biomass. Organic nonfossil material of biological origin. For example, trees and plants are biomass.

Biomass Energy. Energy produced by combusting renewable biomass materials such as wood.

British Thermal Unit (BTU). The amount of energy needed to raise the temperature of one pound of air-free water from 60° to 61°F at a constant pressure of one standard atmosphere; it is found experimentally to be equal to 1054.5 joules.

Carbon Cap. Refers to a suggested regulatory approach of establishing a fixed limit for **emissions** of **carbon dioxide** from all sources in a country or region.

Carbon Dioxide (CO_2). The **greenhouse gas** whose overall impact on trapping heat in the atmosphere is being most affected by human activities. Carbon dioxide also serves as the reference to compare all other greenhouse gases (see **Carbon Dioxide Equivalents**).

Carbon Dioxide Equivalents (CDE). A measure used to compare the **emissions** from various **greenhouse gases** based upon their **global warming** potential (GWP). Carbon dioxide equivalents are commonly expressed as "million metric tons of carbon dioxide equivalents" (MMTCDE) or "million short tons of carbon dioxide equivalents" (MSTCDE). The carbon dioxide equivalent for a gas is derived by multiplying the tons of the gas by the associated GWP.

Carbon Sequestration. A term used to describe any process that removes carbon from the atmosphere. Includes the growth of **biomass** (which converts atmosphere CO_2 into solid, woody carbon), the synthesis of **fossil fuels** (which traps carbon underground as oil, coal, or natural gas, and takes place over geological time scales). These are the only large-scale means of carbon sequestration taking place. Some scientists have suggested the active pumping of CO_2 to the ocean floor as a futuristic, active method of carbon sequestration.

Carbon Sinks. Carbon reservoirs that take in and store more carbon than they release. Carbon sinks can serve to partially offset **greenhouse gas** emissions (**carbon sequestration**). Forests and oceans are common carbon sinks.

Centigrade. A temperature scale used in most countries and by all scientists in which the difference between the freezing point of water and the boiling point of water at sea level is divided into one hundred degrees.

Clayton Antitrust Act. A 1914 amendment to the **Sherman Antitrust Act** prohibiting price discrimination, tying, and exclusive dealing arrangements and mergers, where the effect may be to substantially lessen competition or tend to create a monopoly in any line of commerce. In many circumstances the act also prohibits interlocking directorships involving competitors.

Clean Air Act (CAA). The act was enacted in 1970 and was substantially revised in 1977 and 1991. The original CAA mandated air safe enough to protect the public's health by May 31, 1975 and required the setting of National Ambient Air Quality Standards (NAAQS) for major primary air **pollutants**. The act established the **Environmental Protection Agency (EPA)** and gave the administrator power to establish the standards. The act authorized the EPA to list and regulate various hazardous air pollutants. The 1990 Amendments to the Clean Air Act addressed **emissions** standards for both electric utility and mobile sources (cars and trucks). The amendments phase in tougher standards through the beginning of the next century.

Climate. The average weather (usually taken over a thirty-year time period) for a particular region and time period. Climate is not the same as weather, but is the average pattern of weather for a particular region. Weather describes the short-term state of the atmosphere. Climatic elements include precipitation; temperature; humidity; sunshine; wind velocity; phenomena such as fog, frost, and hailstorms; and other measures of the weather.

Climate Model. A quantitative way of representing the interactions of the atmosphere, oceans, land surface, and ice. Models can range from relatively simple to quite comprehensive. Also see **General Circulation Model**.

Cogeneration. The sequential production of electricity and useful **thermal energy** from the same source. There is heat energy left over from the production of electricity with any known technology, including steam; facilities fueled by fossil, **biomass,** or nuclear energy; reciprocating engines (Otto or **Diesel** cycle); combustion turbines (Brayton cycle) **fuel cells**; or from plants that combine cycles. Power plants which sequentially capture and use this heat are known as cogeneration plants, or **combined heat-and-power (CHP)** plants. See **Trigeneration**.

Combined-Cycle Plants. Any power plant that utilizes more than one basic energy-conversion cycle. The usual meaning is a combined-cycle gas **turbine** (CCGT) which uses a gas turbine (Brayton cycle), then recovers exhaust heat to produce steam used in a steam turbine (**Rankine cycle**). Another example is a **fuel-cell** plant that produces electricity and replaces the combustor of a gas turbine, which is referred to as a fuel-cell gas-turbine hybrid.

Combined Heat and Power (CHP). Another way to describe **cogeneration**, signifying that heat and power are sequentially produced in one process, typically with higher efficiency than separately producing heat and power.

Decarbonize. This term refers to the process of converting the economy to use less fossil carbon. Actions to decarbonize the world economy include increasing the efficiency of power and heat production, increasing the efficiency of converting power and heat to comfort, eliminating energy waste, and increasing the percent of energy produced from nonfossil fuels—nuclear and all renewable energy.

Deforestation. Those practices or processes that result in the change of forested lands to nonforest uses. Deforestation is often cited as one of the major causes of the enhanced **greenhouse effect** for two reasons: (1) the burning or decomposition of the wood releases **carbon dioxide**; and (2) trees that once removed carbon dioxide from the atmosphere by **photosynthesis** are no longer present and contributing to carbon storage.

Demand-Side Management. Deliberate intervention by a utility in the marketplace to reduce or change the time of customer use of electricity. Electric utilities have been required to include demand-side management in the development of long-range planning, including load management, conservation, **peak shaving**, and strategic load building and load shifting. Typically these measures include incentives for participants and/or the utility. Some state regulators are now considering demand-side management for gas utilities.

Department of Energy (DOE). The U. S. federal cabinet-level agency responsible for monitoring and guiding our long-term energy strategy.

Deregulation. The act of freeing from regulation (especially from governmental regulations).

Diesel Engines. A type of internal combustion engine that draws air in an intake stroke as the piston travels to the bottom of the cylinder, whereupon valves are closed and the air is compressed by the returning piston, thus heating the air above the self-ignition temperature of the fuel. Upon completion of the compression stroke, five droplets of fuel are injected into the cylinder and are ignited by the hot air. The resulting burning of the fuel increases the pressure and drives the piston in the power stroke. Named after its inventor, Rudolph Diesel, the diesel engine has a higher thermal efficiency than the spark-plug ignited, or Otto cycle, engine.

Diffusion of Innovations. The comprehensive study of the diffusion of innovations, ranging from the elements of diffusion and the history of diffusion research to developers of innovation, change agents, and the consequences of innovations. Dr. Everett Rodgers summarizes the field in his book *Diffusion of Innovations* (4th ed., The Free Press, 1995), and Dr. Jesse H. Ausubel at the Rockefeller Institute has documented the patterns of diffusion of a wide variety of human actions.

Distributed Power Generation. (1) Distributed power generation is application of any small-scale power-generation technology that provides electric power at a site closer to customers than central station generation, and is usually interconnected directly to the customer's facilities. (2) Generating power at multiple locations instead of at large central locations. The principal reasons for dispersing electric generation are to enable the recovery and use of waste heat and to avoid distribution and transmission losses and costs.

District Energy Systems. District energy systems distribute steam, hot water, and/or chilled water from a central plant to individual buildings through a network of pipes. District energy systems aggregate **thermal energy** loads and make possible the use of larger-scale **com-**

bined heat-and-power technology and enable the use of normally wasted heat from existing electric power plants.

Econometric Models. Computer models that attempt to capture various relationships in the functioning economy and then predict the impact of changing a particular policy.

Electric Power Research Institute (EPRI). Research organization founded in 1972 by electric utilities to improve electric production, distribution, and use. Address: 3412 Hillview Avenue, Palo Alto, CA 95303; (415) 855-2000.

El Niño. A climatic phenomenon occurring irregularly, but generally every three to five years. El Niños often first become evident during the Christmas season (El Niño means Christ child) in the surface oceans of the eastern tropical Pacific Ocean. The phenomenon involves seasonal changes in the direction of the tropical winds over the Pacific and abnormally warm surface ocean temperatures. The changes in the tropics are most intense in the Pacific region; these changes can disrupt weather patterns throughout the tropics and can extend to higher latitudes, especially in Central and North America. The relationship between these events and global weather patterns are currently the subject of much research in order to enhance prediction of seasonal to interannual fluctuations in the climate.

Emissions. The release of a noxious substance (usually a gas when referring to the subject of climate change) into the environment.

Energy Policy Act (EPAct). The Energy Policy Act of 1992 authorizing the **Federal Energy Regulatory Commission (FERC)** to order wholesale **wheeling** of electricity while explicitly restraining its power to order retail wheeling. EPAct also created a new legal category of electricity generating and sales companies called the **Exempt Wholesale Generator (EWG)**, free from **PUHCA** restrictions.

Energy Regulatory Reform and Tax Act (ERRATA). A proposal by the author to modernize regulations and laws impacting energy.

England's National Grid. The set of transmission and distribution wires and related facilities used to distribute electricity throughout Great Britain.

Environmental Protection Agency (EPA). The U. S. cabinet-level agency responsible for protecting the environment.

Exempt Wholesale Generator (EWG). Any person engaged exclusively in the business of owning and/or operating all or part of a facility used for the generation of electric energy exclusively for sale at wholesale or leased to a public utility company.

Fahrenheit. (1) Conforming to the scale used by Gabriel Daniel Fahrenheit in the graduation of his thermometer; of or relating to Fahrenheit's thermometric scale. The Fahrenheit thermometer or scale. It is no longer commonly used anywhere but the United States. (2) Of or relating to a temperature scale that registers the freezing point of water as 32° F and the boiling point as 212° F at one atmosphere of pressure.

Federal Power Commission (FPC). The commission was established when Congress enacted the 1920 Federal Water Power Act and was empowered to collect data concerning the utilization and development of water resources including the location of potential power plant sites. The FPC's responsibilities were expanded under the Public Utility Act of 1935 by giving it the authority to regulate the interstate transmission and sale of wholesale electricity. The 1977 Department of Energy Organization Act established the **Federal Energy Regulatory Commission (FERC),** the successor to the FPC.

Forest Dieback. Refers to a high incidence of decline of forests and individual tree death due to a change in climate conditions or increase in deposition of toxic chemicals from air pollution that makes trees vulnerable to disease and insect predation.

Fossil Fuel. A general term for combustible geologic deposits of carbon in reduced (organic) form and of biological origin, including coal, oil, natural gas, oil shales, and tar sands.

Fossil Fuel Combustion. Burning of any **fossil fuel**, including coal, oil (including gasoline), natural gas, shale oil, tar sands, and peat. This burning, usually to generate energy, combines the carbon in the fossil fuel with oxygen from the air and releases **carbon dioxide** as well as combustion by-products that can include unburned hydrocarbons, **methane**, and carbon monoxide.

Fossil Fuel Efficiency Standard. A measure of the amount of fossil-derived energy used to produce electricity and heat from power plants, proposed herein by the author. The proposed standard measures the energy content of all **fossil fuels** consumed for power generation in **megawatt-hours** and divides this by the megawatt-hours of all electricity produced and of heat recovered from all generation, including fossil fuel, nuclear, and renewable energy. In 1996, 2.07 megawatt-hours of fossil-fuel energy were consumed for every megawatt-hour of total U. S. power produced.

Fuel Cell. An electrochemical cell in which the energy of a reaction between a fuel, such as hydrogen and oxygen, is converted directly and continuously into electrical energy. A fuel cell involves the same chemistry as combustion but converts some of the energy directly to electricity. The main by-products are the same as other combustion, i.e., CO_2 and water.

General Circulation Model (GCM). A global, three-dimensional computer model of the **climate** system which can be used to predict and understand climate. GCMs are highly complex and represent the effects of such factors as reflective and absorptive properties of atmospheric water vapor, **greenhouse gas** concentrations, clouds, annual and daily solar heating, ocean temperatures, and ice boundaries. The most recent GCMs include global representations of the atmosphere, oceans, and land surface.

General Services Administration (GSA). The federal agency responsible for purchasing goods and services for other federal agencies and providing and managing the properties and buildings that house federal agencies.

Global Climate Change (also referred to as Climate Change). The term *global climate change* is sometimes used to refer to all forms of climatic inconsistency, but because the earth's **climate** is never static, the term is more properly used to imply a significant change from one climatic condition to another. In some cases, climate change has been used synonymously with the term *global climate warming*; scientists, however, tend to use the term in the wider sense to include all natural and human-induced changes in climate.

Global Warming. An increase in the near-surface temperature of the earth. Global warming has occurred in the distant past as the result of natural influences, but the term is most often used to refer to the warming predicted to occur as a result of increased **emissions** of **greenhouse gases**. Scientists generally agree that the earth's surface has warmed by about 1 degree **Fahrenheit** in the past 140 years.

Greenhouse Effect. The effect produced by **greenhouse gases** that allow incoming solar radiation to pass through the earth's atmosphere, but reflect some of the outgoing infrared radiation from the solar-heated surface and lower atmosphere back to the surface. This process occurs naturally and has kept the earth's temperature warmer than it would otherwise be. Current life on Earth could not be sustained without the natural greenhouse effect produced by **carbon dioxide** concentrations in the atmosphere. The concern of scientists is that increases in the concentrations of greenhouse gases will raise average temperatures and lead to extreme and violent weather.

Greenhouse Gas. Any gas that absorbs or reflects infrared radiation in the atmosphere. Greenhouse gases include water vapor, **carbon dioxide** (CO_2), **methane** (CH_4), nitrous oxide (N_2O), halogenated fluorocarbons (HCFCs), **ozone** (O_3), perfluorinated carbons (PFCs), and hydrofluorocarbons (HFCs).

Gross Domestic Product (GDP). Measure of the U.S. economy; the total market value of goods and services produced by all citizens and capital during a given period (usually one year).

Gulf Stream. The ocean current which originates in the westward equatorial current, due to the trade winds, is deflected northward by Cape St. Roque through the Gulf of Mexico, where the water is warmed, and then flows parallel to the coast of North America, turning eastward off the island of Nantucket. Its average rate of flow is said to be about two miles an hour. The similar Japan current, or Kuro-Siwo, is sometimes called the Gulf Stream of the Pacific. The warm waters of the Gulf Stream in the North Atlantic release heat to the eastward jet streams that have been cooled in Canada and the Arctic, and thus moderate the northern European climate.

Hydrocarbon. One of a very large group of chemical compounds consisting primarily of carbon and hydrogen; hydrocarbons used for energy production include petroleum crude oil and natural gas.

Ice Core. A cylindrical section of ice removed from a glacier or an ice sheet in order to study **climate** patterns of the past. By performing chemical analyses on the air trapped in the ice, scientists can estimate the percentage of carbon dioxide and other trace gases in the atmosphere at that time.

Infrared Radiation. Electromagnetic radiation of lower frequencies and longer wavelengths than visible light (greater than 0.7 microns [μ.m.]). Solar ultraviolet radiation is absorbed by the earth's surface and reemitted as infrared radiation.

Intergovernmental Panel on Climate Change (IPCC). The IPCC was established jointly by the United Nations Environment Programme and the World Meteorological Organization in 1988. The purpose of the IPCC is to assess information in the scientific and technical literature related to all significant components of the issue of **climate** change. The IPCC draws upon hundreds of the world's expert scientists as authors and thousands as expert reviewers. Leading experts on climate change and environmental, social, and economic sciences from some sixty nations have helped the IPCC to prepare periodic assessments of the scientific underpinnings for understanding global climate change and its consequences. With its capacity for reporting on climate change, its consequences, and the viability of adaptation and mitigation measures, the IPCC is also looked to as the official advisory body to the world's governments on the state of the science of the climate-change issue. For example, the IPCC organized the development of internationally accepted methods for conducting national **greenhouse gas** emission inventories.

Kilowatt (kW). A measure of power output capacity. A unit of power equal to one thousand watts.

Kyoto Protocol. The Kyoto Protocol is the latest step in the ongoing United Nations effort to address **global warming**. The effort began with the United Nations Framework Convention on Climate Change (Convention) signed during the Earth Summit in Rio de Janeiro in 1992. The convention, negotiated in December 1997 in Kyoto, Japan, is intended to stabilize **greenhouse gas** concentrations at a level that will prevent dangerous interference with the global **climate** system. The protocol enters into force when signed by at least fifty-five parties which account for at least 55 percent of the total carbon emissions for 1990.

Megawatt (MW). A measure of power output capacity. One megawatt equals one million watts or 1,000 kilowatts.

Megawatt-hour (MWh). A measure of energy content. One megawatt-hour equals one thousand kilowatt-hours (or 3,412 million **British Thermal Units**).

Methane (CH_4). A hydrocarbon that is a **greenhouse gas** with a **global-warming** potential estimated at 24.5 times a comparable quantity of **carbon dioxide**. Methane is produced through anaerobic (without oxygen) decomposition of waste in landfills, rice paddies, animal digestion, decomposition of animal wastes, production and distribution of natural gas and oil, coal production, and incomplete **fossil fuel** combustion. It is also the primary component of natural gas. The atmospheric concentration of methane has been shown to be increasing at a rate of about 0.6 percent per year and the concentration of about 1.7 parts per million by volume (ppmv) is more than twice its preindustrial value. However, the rate of increase of methane in the atmosphere may be stabilizing.

Monopoly. The exclusive power or privilege of selling a commodity; the exclusive power, right, or privilege of providing some good or service, or of trading in some market; sole command of the traffic in anything, however obtained; as, the proprietor of a patented article is given a monopoly of its sale for a limited time; chartered trading companies have sometimes had a monopoly of trade with remote regions; a combination of traders may get a monopoly of a particular product.

Monopoly Efficiency Gap. A term coined by the author to denote the likely shortfall in efficiency caused by allowing the electric-generation business to enjoy monopoly protection.

Monopoly Franchise. The exclusive rights to generate, distribute, and/or sell a good or service within a specified geographical territory. Such rights are typically granted and enforced by a government agency in order to achieve some social objective.

Municipal Power Companies. Electric-power distribution (sometimes generation as well) companies that are owned by the municipality they serve. These entities are exempt from federal and state income taxes.

National Oceanic and Atmospheric Agency (NOAA). A branch of the U.S. Commerce Department responsible for monitoring weather and the health of our oceans and the atmosphere.

Natural Monopoly. An economic term to describe a condition in which the production and/or distribution and/or sale of a good or service thought to be more efficient if provided by one supplier, using the technology available at the time. Examples are provision of sewers, or electric distribution.

Negawatts. A term coined by Amory Levins to describe the impact on electric production of conservation. For example, new light fixtures that produce the same amount of light in a building with one **megawatt** less peak power are said to produce one "negawatt" of electricity.

Nitrogen Oxides (NOx). Gases consisting of one molecule of nitrogen and varying numbers of oxygen molecules. Nitrogen oxides are the result of combustion of fuel and air (which is roughly 76 percent nitrogen) and are present in the **emissions** of vehicle exhausts and from power stations. In the atmosphere, nitrogen oxides can contribute to formation of photochemical ozone (smog), can impair visibility, and have health consequences; they are thus considered **pollutants**.

Nitrous Oxide (N_2O). A powerful **greenhouse gas** with **global warming** potential of 320. Major sources of nitrous oxide include soil cultivation practices, especially the use of commercial and organic fertilizers, **fossil fuel** combustion, nitric acid production, and **biomass** burning.

Output Standards. Current **EPA** regulations of air **emissions** regulate the concentrations of each **pollutant** in the exhaust gas or as pounds of allowed pollutant per unit of fuel and are thus called *input standards*. By contrast, an output or **performance standard** would regulate the pounds, or kilograms, of each pollutant allowed per unit of useful energy output.

Ozone (O_3). Ozone consists of three atoms of oxygen bonded together in contrast to normal atmospheric oxygen which consist of two atoms of oxygen. Ozone is an important **greenhouse gas** found in both the stratosphere (about 90 percent of the total atmospheric loading) and the troposphere (about 10 percent). Ozone has other effects beyond acting as a greenhouse gas. In the stratosphere, ozone provides a protective layer shielding the earth from ultraviolet radiation and subsequent harmful health effects on humans and the environment. In the troposphere, oxygen molecules in ozone combine with other chemicals and gases (oxidization) to cause smog.

Ozone Layer. A layer in the stratosphere (at approximately twenty miles) that contains a concentration of ozone sufficient to block most ultraviolet radiation from the sun.

Peak Shaving. Generically, the process of eliminating or shifting load from the peak hours of demand. Herein, *peak* refers to any process which eliminates or shifts load on the utility for electricity from the hours of peak demand. Examples include turning off chillers and escalators during peak hours, operating on-site engine generators to produce part of the power, or producing chilling with off-peak power, then storing as cold water or ice to use during the next day's peak demand for electric power.

Performance Standard. An approach to regulating the amount of a **pollutant** allowable per unit of useful output energy.

Photosynthesis. The process by which green plants use light to synthesize organic compounds from **carbon dioxide** and water. In the process oxygen and water are released. Increased levels of carbon dioxide can increase net photosynthesis in some plants.

Photovoltaics (PV). Technology to electronically convert sunlight into electricity.

Pollutants. Strictly, too much of any substance in the wrong place or at the wrong time is a pollutant. More specifically, atmospheric pollution may be defined as the presence of substances in the atmosphere, resulting from man-made activities or from natural processes that cause adverse effects to human health, property, and the environment.

Pollution Offsets. A regulatory allowance that credits the reduction of pollution at some remote location or process to another regulated **emission**-producing process.

Public Utility Holding Company Act (PUHCA). Legislation enacted in 1935 to protect utility stockholders and consumers from financial and economic abuses of utility holding companies. Generally, ownership of 10 percent or more of the voting securities of a public utility subjects a company to extensive regulation under the Securities and Exchange Commission. The Comprehensive National Energy Policy Act of 1992 opened up the power market by granting a class of independent power producers exempt from PUHCA regulation.

Public Utility Regulatory Policies Act (PURPA). One of the five bills signed into law on November 8, 1978 as the National Energy Act. PURPA created various incentives for the development of **cogeneration** facilities that met certain qualifying tests (qualifying facilities or QFs) including the requirements that utilities allow paralleling, provide standby or backup service, and purchase power generated by QFs at the utilities' "full avoided cost." Sections 211 and 212 were added to the Federal Power Act, vesting **FERC** with certain powers to order transmission or "**wheeling**" of electric power under limited circumstances.

Public Utility (Service) Commission. State and federal commissions that regulate the activities of intrastate pipelines and local distribution companies (LDCs), as well as electric, telephone, and water utilities.

Quad. Shorthand for one quadrillion **BTUs**.

Quadrillion. A quantity of energy equal to one quadrillion **British Thermal Units** (BTUs), which is commonly referred to as a "quad." The current world consumption of energy for all uses and from all sources is roughly 310 quadrillion BTUs per year.

R&D. Research and development for new products and processes.

Rankine Cycle. A power cycle that boils water or other fluids to steam at above-atmospheric pressures and then uses that steam or vapor to drive a **turbine** which exhausts vapors at a lower pressure. More specifically, a thermodynamic cycle consisting of heat addition at constant pressure, theoretically isentropic expansion, heat rejection at constant pressure, and theoretically isentropic compression; used as an ideal standard for the performance of heat-engine and heat-pump installations operating with a condensable vapor as the working fluid, such as a steam power plant. Also known as *steam cycle*.

Rate Commission. Refers to the **Public Utility Commission** or body empowered to set prices and rates and to control the profits of the monopoly or regulated utility.

Renewable Energy Portfolio Standards. Requirements in some state and in several versions of proposed U.S. federal electric deregulation legislation that a specified portion of electric power sold by each power company be produced from renewable energy such as **biomass**, solar, wind, and in some definitions, hydroelectric sources.

Renewable Resources. Energy sources that use renewable energy including hydro, wind, solar energy, and geothermal energy, as well as some combustible materials, such as landfill gas, **biomass**, and municipal solid waste.

Rural Electric Agencies. The local electric distribution organizations originally established to electrify rural areas. REAs are exempt from federal income tax and are empowered to borrow money under terms guaranteed by the U.S. federal government.

Rural Electrification Act. The Rural Electrification Act became law in 1936 and was designed to speed the electrification of rural areas by providing access to funds at U.S. federal government interest rates.

Sherman Act. The basic federal antitrust law. Enacted in 1890, it very broadly prohibits unreasonable restraints of trade, including price fixing, **monopolies**, and attempts and conspiracies to create monopolies. It was subsequently strengthened in 1914 by the **Clayton Antitrust Act**.

Stranded Costs. Generically, costs not yet amortized of any utility asset which can no longer be assigned to customers after deregulation or reduction of **monopoly** protection of the utility in question. Herein, stranded costs refer primarily to the unamortized costs of electric-power generation plants built under full monopoly protection, which are not expected to be covered under market competition, and the contracts for power purchase from independent power producers which are above current market prices.

Sustainable Development. A broad concept referring to the need to balance the satisfaction of near-term interests with the protection of the interests of future generations, including their interests in a safe and healthy environment. As expressed by the 1987 United Nations World Commission on Environment and Development (the "Brundtland Commission"), sustainable development "meets the needs of the present without compromising the ability of future generations to meet their needs."

Tax Depreciation. The rules regulating the portion of an asset's original cost that can be depreciated or expensed each year for purposes of calculating taxable income and thus taxes due.

Tennessee Valley Authority (TVA). A federal agency formed in 1933 to manage flood control and produce power in the Tennessee Valley and other defined portions of the U.S. South. The TVA today operates hydro-electric, nuclear, and fossil-based electric generation producing 140 billion kilowatt-hours of electricity per year.

Thermal Energy. Energy used to heat or cool spaces or processes. Examples include hot water or steam used to heat buildings or heat processes, or chilled water used to cool buildings.

Title IV of the Clean Air Act Amendments of 1990. Title IV set regulations for the electric utility industry to reduce annual sulfur dioxide (SO_2) **emissions** by 10 million tons and annual nitrogen oxides (NOx) emissions by 2.0 million tons from 1990 levels by the year 2000. Beginning in the year 2000, total utility SO_2 emissions are then limited to 8.9 million tons and total industrial SO_2 emissions are expected to be 5.6 million tons. Title IV's control of SO_2 emissions instituted two important innovations in U.S. environmental policy. First, it introduced the SO_2 emissions trading program where firms are given permits to release a specified number of tons of SO_2. The government issues only a limited number of permits consistent with the desired level of emissions. The owners of the permits may keep them and release the pollutant, or reduce their emissions and sell the permits. The fact that the permits have value as an item to be sold or traded gives the owner an incentive to reduce their emissions. Second, Title IV established an average annual cap on aggregate emissions by electric utilities. This cap was set at about one-half of the amount emitted in 1980. The emissions cap represents a guarantee that emissions will not increase with economic growth.

Tragedy of the Commons. A term coined by Garret Harden ("Tragedy of the Commons," *Science* 162 [1968]) referring to the generic notion that individuals acting logically in their self-interest can cause damage to the natural processes they depend upon, resulting in a loss of welfare for everyone.

Trigeneration. A term coined by the author in 1986, trigeneration is the conversion of a single fuel into three useful energy products: electricity, steam, or hot water, and chilled water with lower pollution and greater efficiency than is possible by producing the three products separately.

Turbine. A device used to convert the energy in a liquid, or gas, to mechanical power. The energy may be the potential energy of falling water, the chemical energy of natural gas, or the **thermal energy** of steam. These fluids give up their energy to rotate a wheel on a shaft. The rotating shaft may be used directly for energy as in an automobile, or converted to electrical energy in a generator.

Uniform Generation Performance Standard. See **Performance Standard**.

Vostock Base. A Russian base in the Antarctic where there have been new ice layers formed from snow every year and no annual melting over a period in excess of 160,000 years. The **ice cores** taken at Vostock thus contain a record of the chemical composition of ice from each past year, from which scientists have deduced information about the climate over the past 160,000 years.

Vostock Ice-Core Data. Data from **ice cores** taken at the Vostock base maintained by Russia on the Antarctic continent.

Wheeling Power. The movement of power from any generator source over wires owned by another entity to the ultimate consumer.

White House Council on Environmental Quality (CEQ). The council was established in 1969 to advise and assist the president in the development of environmental policies and proposed legislation as requested by the president and to act as a clearinghouse of environmental information for the executive office of the president.

Xonon Combustor. A trademark name for a series of combustors employing catalytic technology that causes the fuel to combust at temperatures below the temperatures at which oxides of nitrogen are formed. The developer and owner of the technology is Catalytica, a NASDAQ-listed public company.

Index

Achilles, 71
acid rain, 1,3, 7, 54, 72, 79, 112, 131
action, recommendations for, 132–33, 134,
    136, 161–62, 164, 165, 169, 171–72,
    173, 175, 176, 177, 179, 182, 184,
    186, 188, 189, 198, 230–34
actuarial probability, 170–72
age of power plants, 181
Alliance to Save Energy, 206, 209, 210
Antarctic, 18, 76
antitrust statutes, 27, 28, 53
    See also Clayton Antitrust Act; Sherman
    Antitrust Act
Apple Computers, 67–68
Argentina, 11, 27, 32, 78, 117
Arrhenius, Svante, 15
Atlantic City, N. J., 174–75
atmosphere, chemical composition, 74
Atmospheric Pollution Prevention Divi-
    sion, 92
automobiles. See vehicles

backup power, 169, 172
Baltimore, Md., 193–94

Baltimore district steam system, 204–205
Bayless, Charles, 42, 62, 203, 218
Becker, Daniel F., 122
Bell, Alexander Graham, 29
Bethlehem Steel, 85
biomass, 6, 10, 79, 143, 222
Boston Edison, 113
Bradford, Peter, 227–28
Britain. See United Kingdom
British Petroleum, 79
building codes, 132–33, 136
bundling, 26, 172–73
Bundy, McGeorge, 71
Busch, Adolphus, 100

cable television, 31
California, 179, 180, 226
carbon caps, 124, 125, 206, 207–209, 213,
    217, 218, 220
carbon dioxide
    atmospheric, 2, 10, 15, 17, 19, 74,
    75–76, 77, 119–29, 205, 208
    emissions, 3, 5, 15, 16 (chart), 20, 23,
    25, 55, 75–76, 79, 83, 117, 124, 225

(chart), 226
reduction of, 4, 8–9, 11, 33, 69, 84, 85, 105–18, 127–29, 130, 145, 146, 152, 154, 182, 196, 205–206, 209, 211, 216, 219, 224, 228
carbon monoxide, 65, 73, 185
Carnegie, Andrew, 26
Casten, Thomas R., 63, 216, 218, 227
Catalytica, 65, 118
Caterpillar, 63
Cherry Hill Housing Project, 189
Chile, 11, 27, 32
China, 18, 218
Cicio, Paul, 85
Cinergy, 61, 194
Clayton Antitrust Act, 26, 172
Clean Air Act of 1972, 4, 112, 117, 177, 182–83, 209, 220
climate change, global, 2, 3, 7, 22, 67, 75, 84, 123, 206, 224, 228
climate models, 15, 16–17, 18
Clinton, William, 22, 221
clouds, 75
CMS Company, 61
coal, 121, 221, 222
coal-fired power plants, 46, 52
cogeneration (combined heat and power [CHP]), 5, 33, 49, 53, 63–65, 68, 85, 138, 148, 155–57, 163–64, 167–68, 173–75, 183, 186, 187, 192, 209
Cogeneration Development Corporation, 10
Colorado, 164, 211, 219
Coors Brewery, 171, 219
combined heat-and-power plants. See cogeneration
Common Purpose, 216
Compaq, 153
competition, 1, 5, 41, 67–69, 82, 106, 113–14, 116–17, 131, 145, 147, 149, 151, 161, 164, 165, 172, 204–205, 207, 217, 219, 224, 228–29, 234
Comprehensive Electricity Competition Plan, 207, 211

computer-chip technology, 67–68
conservation of energy, 4, 5, 6, 101, 127, 130–34, 142–43, 196, 215, 232–33
Consolidated Edison, 63
Constitution, U.S., 107, 108, 199
consumption of energy by humans, 10
cooperation vs. competition, 53–55
Cooperative Institute for Research in Environmental Sciences, 122
cost of electricity. See rates, electric
Cottrell, Fred, 74
Council of Economic Advisors, 84, 90, 146, 147
Council on Environmental Quality, 84
Cummins Cogeneration Company, 10, 63, 167
Cummins Engine Company, 3, 54, 63, 100
Cummins Tech Center, 3
deforestation, 54, 71, 76, 131
Dell, Michael, 153
demand side management (DSM), 90, 96, 97
Department of Energy (DOE), 92, 109, 216
Department of Renewables and Efficiency, 216
Department of Transportation, 144
depletion allowances, 200
depreciation, 187–88
deregulation, 27, 62, 69, 78, 105–18, 164, 165, 196, 198, 201–205, 206, 211, 212–13, 217, 219, 228, 230
See also Argentina; Chile; telephone, deregulation of; United Kingdom
Detroit Diesel, 63
diesel engines, 54, 63, 99
diseases, 71, 72
district energy systems, 134–37, 158–59
Dow Chemical Company, 85
Drake, Edwin Laurentine, 36
Drucker, Peter, 223–24
Duke Power Corporation, 61
Dupont, 85

Dying of the Trees, The, 3
Dyson, Freeman, 79

Earth Day, 20
Eastman Kodak, 27
economic models, 10, 23, 117, 146, 195–97
Economist, 195
economy, U.S., 23
Edison, Thomas, 5, 37, 40, 45
Edison Electric Institute, 53
efficiency
  barriers to, 148–94, 198–200, 205, 213, 228
  of energy production, 2, 4, 5, 6, 7, 8, 10–11, 23, 25, 41, 48–54, 60, 63, 66 (chart), 67, 106, 125, 126, 128, 131, 135, 147–97, 200, 209, 210, 212, 213–19, 223, 226, 231
  of energy use, 5, 77, 86–87, 98–101, 103, 130, 138–40, 144–45, 147–97, 208, 231
electricity distribution systems, 165
electric industry, development of, 32–33
electricity, price of. *See* rates, electric
Electric Power Research Institute, 33
Electric Power Supply Association, 112
El Niño, 2, 17, 18
End of Liberalism, The, 30
Energy and Society, 74
Energy Policy, 123
Energy Policy Act of 1992 (EPACT), 44, 162, 163
energy-production technologies, 7
Energy Regulatory Reform and Tax Act (ERRATA), 230–34
Energy Star designation, 92
Entergy, 61
Environmental Protection Agency (EPA), 92, 152, 179
environmental regulations, 7
environmentalists, 7
ethanol, 143

Falcone, Joseph, 218
fear, motivation of, 55, 106
Fechter, Gary, 91
Federal Energy Regulatory Commission (FERC), 211
Federal Power Act (FPA), 44
Florida, 18
forest dieback, 3
Forrister, Dirk, 84
Fossil Fuel Efficiency Standard (FFES), 8–9, 109, 114, 128, 206, 213–19
fossil fuels, 6, 22, 110, 213, 215 (chart), 219
France Telecom, 31
free markets, 4, 11, 81, 196, 204, 212, 213, 217
Freudenthal, Peter, 64

gas. *See* natural gas
gas-turbine power generation, 65, 118 (chart), 150, 151, 156, 187
Gelbspan, Ross, 2, 73
General Electric, 118
General Services Administration (GSA), 139
Georgia Pacific, 85
geothermal power, 10
Ghamarian, Ahmad, 218
Global Climate Coalition, 79
global warming, 17, 18, 22, 72, 76, 79, 143
Golden, Colo., 219
Gould, Jay, 26
greed, 55, 59
greenhouse effect, 2, 15
greenhouse gases, 8, 17, 23, 33, 76, 78, 131, 145
gross domestic product (GDP)
  energy consumption per (chart), 142
Guatemala, 154
Gulf Stream, 75

Haldane, J. B. S., 79, 80
Hall, Mark, 216, 218
Hardin, Garrett, 80, 95, 227

heat, 4, 57, 208, 210
   electric, 173
Heat Is On, The, 2, 73
Hogan, Kathleen, 92
Hopkins, John D., 88
House Subcommittee on Energy and
   Power, 211
Human Action, 90
Hunt, Sally, 105
hydropower, 6, 49

ice ages, 20
Iliad, 70–71
Illinova Power, 62, 203
Imagined Worlds, 79
independent power producers (IPPs), 59,
   61, 111, 153
India, 31, 218, 227
Industrial Revolution, 15, 17
infrared radiation, 15, 75
innovations, diffusion of, 33–36, 54, 219
input-based standard, 178
Insull, Samuel, 40, 58, 60
interconnections, 165–69, 170
International Business Machine (IBM),
   67–68
interstate commerce, 199
Interstate Commerce Commission, 27
investor-owned electric utility companies,
   59–62, 220

Japan, 98, 101, 155, 219
Jefferies, David, 105

Klein, Milton, 218
Klein, Richard M., 76
Kyoto Protocol of 1997, 8–9, 77, 78–79,
   124, 126, 207, 218, 219, 226

Little, Charles E., 3, 76–77
lobbying, 54
London, Ontario, 168–69
Long Island Lighting Company (LILCO),
   58–59

Long Island Power Authority (LIPA),
   41–42, 164
Louisiana, 194
Louisiana Purchase, 10
Louisville Gas and Electric, 61
Lovins, Amory and Hunter, 89, 97, 195
lowest achievable emission rate (LAER),
   118
Lowi, Theodore J., 30
Lynd, Lee, 115

Marble Hill nuclear power plant, 61
Maryland, 194
Massachusetts, 112, 164, 179, 180, 190–
   91, 226
Massachusetts Department of Public Utilities, 192
Massachusetts Institute of Technology
   (MIT), 20, 191–93
Matsukata, Shichiro, 98
McDonald's, 102
mercantilism, 108
Mercer Medical Center, 185
monopoly efficiency cap, 226
monopoly regulations, 117, 210, 219
monopoly rents, 41, 63
monopolies, 4, 6, 25–69, 165, 172, 203
Morgan, J. P., 26
municipal electric companies, 163, 164,
   202, 203

National Ambient Air Quality Standard,
   179
National Economic Research Associates
   (NERA), 105–106
National Marketplace for the Environment,
   119
National Oceanic and Atmospheric Agency
   (NOAA), 17, 122
National Snow and Ice Data Center, 122
Nation's Energy Corporation, 171
natural gas, 83, 121, 148, 151, 155–56,
   219, 221
natural monopolies, 27, 29, 31, 33, 40, 44,

56, 98, 158, 164, 217
negawatts, 89, 97
Nelson, Bob, 160
Nepal, 31
Newcomen, Thomas, 74
New England Electric System (NEES), 112–13
New Hampshire Public Service (PSNH), 62
New Jersey, 179
New Jersey Board of Public Utilities, 175
New Source Performance Standards (NSPS), 179
New York City, 5, 45
New York Earth Day Conference, 92
New York Times Magazine, 122
Niagara Falls, N.Y., 37–38
nitrous oxide emissions (NOx), 3, 11, 65, 112, 117–18, 177–82, 186, 220, 224
noblesse oblige, 70
North Carolina, 194
Norway, 27, 32
nuclear power, 41, 49, 114, 150, 151

ocean currents, 75
oil, 151, 221
Ontario Hydro, 41, 168–69
Organization of Petroleum Exporting Countries (OPEC), 99, 141, 157, 210, 221, 229
Otto cycle engines, 99–100
outage, electrical, 171
output-based standard, 179

Palmer, Fredrick, 20
parallel processing, 80–81
Pearl Street generating plant (Manhattan), 5
peat bogs, 121
Peña, Federico, 216, 234
petroleum industry, 229
Philadelphia, 138, 160
photosynthesis, 120
photovoltaic energy, 79, 114
Pignatelli, Jim, 61

polar ice caps, 76
pollution, 3, 8, 11, 77
    See *also* acid rain; carbon dioxide; nitrous oxide; sulfur
pollution-control technologies, 5
pollution limits, 111
population, world, 21, 71, 75, 79, 126, 206
power plants, age of, 3
Pratt & Whitney Canada, 118
Price It Right, 209
price of electricity. See rates, electric
Prize, The, 229
"Public Power—Time's Up," 203
Public Utility Holding Company Act of 1935 (PUHCA), 44, 55–56, 57–59
Public Utility Regulatory Policies Act of 1978 (PURPA), 27, 32, 44, 56, 59–61, 111, 154, 167–68

rainfall, 18
Rankine, William J. M., 46
Rankine-cycle generators, 46, 49, 53, 111, 149, 150
rates, electric, 5, 11, 69, 163, 212, 216, 224, 228
Raymo, Maureen, 20
real-time pricing (RTP), 175–76
Reddy Kilowatt, 54
regulation of energy markets, 4, 28, 158–59, 168
regulation of energy production, 132, 158–59, 185–86
Reike, Bob, 160
Reikins, Timothy D., 76
renewable energy sources, 22, 114, 130, 206, 211, 213, 214
    See *also* biomass; geothermal power; hydropower; nuclear power; solar power; wind power
    Rio de Janeiro. See United Nations Climate Conference
robber barons, 26
Rockefeller, John D., 26, 36–37
Rocky Mountain Institute, 89

Rogers, Jim, 61–62
Romm, Joseph, 216
Roy, Ellen, 112
Royal Dutch Shell, 79
Rural Electric Agencies, 7
Rural Electrification Act, 202
Rural Electrification Administration (REA). See Rural Utilities Service
Rural Electrification Associations (REAs), 108, 220
Rural Utilities Service (RUS), 202

Salt River, 108
Saw Mill River Courts, 167–68, 169
Scambos, Ted, 122
Schaefer, Dan, 211
Schwartz, Bertram, 63, 64
Scientific American, 36
scientific method, 21, 72
SCONOx, 118
self-generation, 56
self-interest, 81, 85, 106, 132
Sematech, 54
Sherman Antitrust Act, 26, 172
Shoreham nuclear power plant, 41, 58–59
Sierra Club, 122
small-scale generation, 57
Smith, Adam, 81, 82
Smithsonian Museum, 49–50
spare generating capacity, 170
Stalone, Charles, 218
standard of living, 2, 9, 23, 206
steam distribution, regulation of, 158, 160
steam-turbine technology, 45–53
solar power, 10
Southern California Edison, 61
standard of living, 77, 130, 205, 212
steam distribution, 160
steam engine, 74
Stern, Todd, 84
subsidies
    consumer, 6, 203–204
    fossil fuel, 6, 127, 140–43, 198, 200
sulfur, 3, 17, 112, 177–79, 207, 220, 224

sustainable energy future, 142–43, 213
Sweden, 27, 32

taxes, 6, 204–205, 209–11
technology. *See* energy-production technologies; pollution-control technologies; steam-turbine technology
telephone, deregulation of, 30, 31
temperature changes, 19 (chart), 74, 76
Tennessee Valley Authority (TVA), 7, 108, 203, 220
Tesla, Nikola, 37–38
Texas, 18
Thailand, 71
Thatcher, Margaret, 11, 27, 105–106
thermal energy. *See* heat
Thompson, Mike, 85–86
Titusville, Pa., 36
"Tragedy of the Commons," 80, 95, 130, 132, 226, 228
transition payments, 112
Trenton District Energy Company, 184–85
Trigen Energy Corporation, 10, 52, 62, 88, 139, 146, 152, 160, 171, 174–76, 193–94, 204, 216, 218, 226
Tucson Electric Power, 62, 171
Tulsa, Okla., 175

"Uniform Generation Performance Standard," 112
UniSource, 62, 171
United Kingdom, deregulation in, 11, 32, 78, 105–107, 117, 152, 196, 217, 219–20
United Nations, 1
United Nations Climate Conference of 1992, 16
United Nations Climate Conference of 1995, 17
University of Colorado, 220
U.S. Generating Company, 112–13

vehicles, 56
Von Hayek, Fredrich A., 82

Von Mises, Ludwig, 90–91, 93, 99
Vostock ice-core data, 18, 120

Washington Post, 218
waste of energy, 5, 6, 11, 23, 205, 212, 214, 219
waste of fuel, 16, 101, 219
Watt, James, 74
Wealth of Nations, 81
weather, 2, 75, 122
Weiser, Mike, 63
Western Fuels Association, 20
Weyerhaeuser, 85

Whirlpool, 85
White House Task Force on Climate Change, 84
wind power, 114
Wirth, Timothy, 119–20, 123, 218
Woods Hole Oceanographic Institute, 20
Wyoming, 222

XONON combustor, 118

Yellen, Janet, 84, 90, 146, 147
Yergin, Daniel, 229
Yohe, Gary, 195